Nanotechnology for Environmental Remediation

Modern Inorganic Chemistry

Series Editor: John P. Fackler, Jr., *Texas A&M University*

Sung Hee Joo
I. Francis Cheng

Nanotechnology for Environmental Remediation

With 79 Illustrations

 Springer

Sung Hee Joo
Environmental Engineering Program
Civil Engineering Department
Auburn University
Auburn, AL 36849
USA
joosung@auburn.edu

I. Francis Cheng
Department of Chemistry
University of Idaho
Moscow, ID 83844
USA
ifcheng@uidaho.edu

ISBN: 978-1-4419-2112-3   e-ISBN: 0-387-28826-0
                              e-ISBN: 978-0-387-28826-0
Printed on acid-free paper.

9 8 7 6 5 4 3 2 1

springer.com

Preface

The book covers the recently discovered oxidative process driven by zero-valent iron (ZVI) in the presence of oxygen and a further developed system which is named ZEA (Zero-valent iron, EDTA, Air). Future potential applications for environmental remediation using this process are also discussed. The oxidative process was discovered during the course of molinate (a thiocarbamate herbicide) degradation experiments. Both ferrous iron and superoxide (or, at pH < 4.8, hydroperoxy) radicals appear to be generated on corrosion of the ZVI with resultant production of strongly oxidizing entities capable of degrading the trace contaminant. Fenton oxidation and oxidative by-products were observed during nanosized ZVI (nZVI)-mediated degradation of molinate under aerobic conditions. To assess the potential application of nZVI for oxidative transformation of organic contaminants, the conversion of benzoic acid (BA) to p-hydroxybenzoic acid (p-HBA) was used as a probe reaction. When nZVI was added to BA-containing water, an initial pulse of p-HBA was detected during the first 30 minutes, followed by the slow generation of additional p-HBA over periods of at least 24 hours. The ZEA system showed that chlorinated phenols, organophosphorus and EDTA have been degraded. The mechanism by which the ZEA reaction proceeds is hypothesized to be through reactive oxygen intermediates. The ZVI-mediated oxidation and ZEA system may be useful for in situ applications of nZVI particles and may also provide a means of oxidizing organic contaminants in granular ZVI-containing permeable reactive barriers.

The purpose of this book is to provide information on the recently discovered chemical process, which could revolutionize the treatment of pesticides and contaminated water. It also aims to offer significant insights to the knowledge for potential applications of ZVI-based technology.

Oxidative degradation of herbicides (e.g., molinate) with its pathway, mechanistic interpretation of the data, modelling/simulation, implication for remediation applications, experimental methodology suitable for pesticides analysis, and ZEA (Zero-valent iron, EDTA, and Air) system with its degradation mechanism are included.

Acknowledgments

We would like to deeply acknowledge Dr. Christina Noradoun who contributed her expertise in Chapter 5. Dr. Joo wishes to thank former mentors, Professor David Waite and Dr. Andrew Feitz for their advice during research on this topic. Special thanks go to Dr. Joseph Pignatello who provided insightful comments in preparing the manuscript.

Finally we would like to appreciate reviewers' comments, which improve the quality of this book and the senior editor, Kenneth Howell who supported us in the preparation of this book.

Contents

Abbreviations and Symbols

ZVI   Zero Valent Iron

ZEA   Zero-valent iron, EDTA, and Air

PRB   Permeable Reactive Barrier

SPME   Solid-Phase MicroExtraction

GC   Gas Chromatography

MS   Mass Spectrometer

HPLC   High Performance Liquid Chromatography

ELISA Enzyme Linked Immuno Sorbent Assays

SOD   Superoxide Dismutase

SEM   Scanning Electron Microscope

TEM   Transmission Electron Microscope

XRD   X-Ray Diffraction

XPS   X-ray Photoelectron Spectroscopy

1
Introduction

Nanotechnology, which is a growing and cutting edge of chemistry, has been of considerable interest in the multidisciplinary research area including chemistry, biochemistry, medicine, and material science. Nanoscale materials have received significant interest; in particular, nanoscale zero-valent iron (denoted here as nZVI) has been attractive for environmental remediation as it is nontoxic, abundant, and potentially least costly. The use of nZVI for remediation provides fundamental research opportunities and technological applications in environmental engineering and science. Zero-valent iron (ZVI) has proven to be useful for reductively transforming or degrading numerous types of organic and inorganic environmental contaminants.

Few studies, however, have investigated the oxidation potential of ZVI. The recently discovered ZVI *oxidative* process and the further modified process in the presence of ethylenetetraaminediacetic acid (EDTA) are described, and the potential future applications are discussed. The discovered reaction processes can be widely used to treat pesticides, herbicides, and industrial chemicals and purify contaminated water for domestic use. One of the most interesting, and potentially least costly, methods for their degradation involves the use of elemental iron (Fe(0)). While Fe(0) or ZVI has been used principally to degrade contaminants in subsurface environments by placing ZVI barriers across the groundwater flowpath, the possibility also exists of using particulate ZVI, which could be either pumped into a contaminated aquifer or dispersed through contaminated sediments.

The focus of the work reported here is on the degradation of agrochemicals, which are widely used worldwide and yet for which low-cost treatment is scarce. Organic compounds such as herbicides, pesticides, and insecticides are of considerable concern with respect to contamination of waters and sediments in the environment and, where inappropriate deposition has occurred, must be removed or degraded. Pesticide contamination of surface waters, groundwater, and soils due to their extensive application in agriculture is a growing, worldwide concern. Pesticides affect aquatic ecosystems and accumulate in the human body. In many countries, the presence of agrochemicals in drinking water supplies is of particular concern, and there is a genuine need for efficient and cost-effective remedial technologies. Thus, the investigation of remediation technology for polluted waters containing trace amounts of herbicides is of environmental interest.

Although there have been approaches to the treatment of pesticide-contaminated soils and waters, ranging from conventional methods such as incineration, phytoremediation, and photochemical processes to innovative methods such as ultrasound-promoted remediation and other advanced oxidation processes, recent studies have shown that many pesticides are susceptible to degradation using ZVI. There have been suggestions recently that use of nZVI could render such an approach particularly attractive because of the high degradation rates that might ensue. Given that many agrochemicals are strongly hydrophobic, use of nanosized ZVI could also facilitate degradation of contaminants sorbed to natural particulate matter. While the use of nZVI appears to be attractive, many questions remain concerning the mode of degradation of dissolved or sorbed contaminants, the effect of solution and surface conditions, and the overall viability of the method.

Finally it would please us greatly if the newly discovered advanced oxidation technology, which is reported in this book, can contribute to advancing science and technology and serve valuable information to all readers (researchers, scientists, engineers, students) in this field for their further research and studies.

1.1. Objectives

The first objective of the work reported here is the examination of the suitability of nZVI to degradation of organic contaminants for the purpose of developing a cost-effective treatment technology.

A second objective of this study is the identification of by-products produced from the ZVI-mediated degradation process of particular contaminants. Any process which generates by-products that are potentially more harmful than the starting material is clearly of limited value. Additionally, identification of any by-products formed may provide insight into the reaction mechanism and suggest approaches by which the technology can be further refined.

The third objective is to clarify the reaction mechanism by which ZVI degrades a chosen contaminant. As noted above, identification of specific by-products may assist in elucidating the mechanism. Other methods, including use of specific probe molecules, examination of the degradation process under varying reaction conditions, and measurement of any reactive transient involved in the degradation process, may assist in this task.

The fourth objective is to assess how the ZVI-based technology may be applied in complex, natural systems and to assess limitations to implementation and the possible avenues for further research that might improve the viability of the process.

Finally the study aims to develop and refine a green oxidation system capable of degrading key priority pollutants (or xenobiotics).

1.2. Outlines

A review of literature relevant to the subject area (ZVI, pesticides contamination and treatment, management practices) is presented in Chapter 2. Firstly, the degradation of organic compounds by granular ZVI in permeable reactive barriers (PRBs), by ZVI colloids, and by nanosized ZVI is described. Secondly, chemical characteristics and environmental impacts of pesticides are described, and common treatment techniques (e.g., incineration, photochemical processes, bioremediation.) are presented and compared with the ZVI technology. Thirdly, preliminary results of screening studies used to assess the applicability of nZVI for treatment of organochlorine insecticides, herbicides, and organophosphate insecticides are presented.

In Chapter 3, materials and analytical techniques that were used in the experimental program are described. In particular, the method for synthesizing nZVI particles is presented, as are the techniques used to characterize the material produced. The methods used to quantify both the starting material as well as organic and inorganic intermediates and products are also outlined, including solid-phase microextraction (SPME) GC/MS, high-performance liquid chromatograpy (HPLC), and colorimetric methods for Fe(II) and H_2O_2 analysis.

Results of screening studies showed that the thiocarbamate herbicide S-ethyl perhydroazepin-1-carbothioate, commonly known as molinate, is particularly susceptible to degradation by ZVI. This compound is widely used in rice-growing areas worldwide and represents a significant water quality problem. In light of these factors, detailed studies of the degradation of molinate by nanosized ZVI have been undertaken, and results are presented in Chapter 4. The results of these studies suggest that molinate is degraded by ZVI via an oxidative process if oxygen is present. The effects of oxygen, pH, and systems conditions on generation of key intermediates (ferrous iron and hydrogen peroxide) are reported in this chapter.

As a further development of the process driven by ZVI, a system, which is named ZEA (for its components Zero-valent iron, EDTA, and Air), is defined, and xenobiotic degradation by the ZEA system with the oxidation mechanism is described in Chapter 5. The mode of oxidative degradation of organic compounds by nZVI is investigated in more detail in Chapter 6, where results of studies on the degradation of benzoic acid by nZVI are reported. Quantification of the oxidative capacity of the technique under specific system conditions is provided in this chapter, as is the importance (or lack thereof) of heterogeneous versus homogeneous processes. In Chapter 7, further investigation on the effectiveness of ZVI for degradation of contaminants of particular concern in drinking waters and recycled wastewaters in continuous column studies is reported. In addition, the experimental results reported in the previous chapters are summarized, and conclusions of this research are presented. Further research needs are described, as are the possibilities for application of nZVI-based technologies.

2
Literature Review

2.1. Zero-Valent Iron (ZVI)

2.1.1. Iron Use in the Environment

Iron is one of the most abundant metals on earth, making up about 5% of the Earth's crust, and is essential for life to all organisms, except for a few bacteria (LANL, 2005). It has been recently recognized as one of the most important nutrients for phytoplankton. For example, as a potential strategy to reduce global warming, scientists have been interested in fertilizing iron in ocean. Adding iron into high-nutrient, low-chlorophyll (HNLC) seawaters can increase phytoplankton production and export organic carbon, and hence increase carbon sink of anthropogenic CO_2, to reduce global warming (Song, 2003). The addition of relatively small amounts of iron to certain ocean regions may lead to a large increase in carbon sequestration at a relatively low financial cost (Buesseler and Boyd, 2003).

One of the most exciting and fastest growing areas of scientific research is the use of nanoscale ZVI for environmental remediation. Zero-valence state metals (such as Fe^0, Zn^0, Sn^0, and Al^0) are surprisingly effective agents for the remediation of contaminated groundwaters (Powell et al., 1995; Warren et al., 1995). ZVI is the preferred and most widely used zero-valent metal because it is readily available, inexpensive, and nontoxic (Gillham and O'Hannesin, 1994; Liang et al., 2000). ZVI (or Fe^0) in particular has been the subject of numerous studies over the last 10 years. ZVI is effective for the reduction of a diverse range of contaminants, including dechlorination of chlorinated solvents in contaminated groundwater (Matheson and Tratnyek, 1994; Powell et al., 1995), reduction of nitrate to atmospheric N_2 (Chew and Zhang, 1999; Choe et al., 2000; Rahman and Agrawal, 1997), immobilization of numerous inorganic cations and anions (Charlet et al., 1998; Lackovic et al., 2000; Morrison et al., 2002; Powell et al., 1995; 1999; Pratt et al., 1997; Puls et al.; Su and Puls, 2001), reduction of metallic elements (Morrison et al., 2002), and the reduction of aromatic azo dye compounds (Cao et al., 1999; Nam and Tratnyek, 2000) and other organics such as pentachlorophenol (Kim and Carraway, 2000) and haloacetic acids (Hozalski et al., 2001).

The reduction process in ZVI systems is a redox reaction where the metal serves as an electron donor for the reduction of oxidized species. Under anaerobic conditions, and in the absence of any competitors, iron can slowly reduce water resulting in the formation of hydrogen gas (Tratnyek et al., 2003), i.e.

$$Fe^0 + 2H_2O \rightarrow Fe^{2+} + H_2 + 2OH^- \tag{2.1}$$

Other reactants may also be reduced by iron. For example, the overall surface-controlled hydrogenolysis of alkyl chlorides (R-Cl) by Fe^0 is likely to occur as follows (Kaplan et al., 1996; Matheson and Tratnyek, 1994; Tratnyek et al., 2003):

$$Fe^0 + R\text{-}Cl + H^+ \rightarrow Fe^{2+} + R\text{-}H + Cl^- \tag{2.2}$$

A schematic of the reduction of tetrachloroethene is given in Figure 2.1a,b. Figure 2.1a shows perchloroethylene (PCE) reacting on the surface of ZVI (Zhang et al., 1998), where ZVI is oxidized to Fe(II) while PCE is dechlorinated. Boronina et al. (1995) studied organohalides removal using metal particles such as magnesium, tin, and zinc and observed that the ability of Zn and Sn particles to decompose the chlorocarbons depends on the quantity of metal and its surface properties and increased in the following order: Sn (mossy) < Sn (granular) < Sn (cryo-particles) < Zn (dust) < Zn (cryo-particles).

The destruction of pesticides using ZVI is also possible. The reductive dechlorination of alachlor and metolachlor (Eykholt and Davenport, 1998) and reductive dechlorination and dealkylation of s-triazine (Ghauch and Suptil, 2000) were observed in laboratory studies. Ghauch (2001) even found rapid removal of some pesticides (benomyl, picloram, and dicamba) under aerobic conditions (8 ppm DO) ($\tau_{1/2}$ of a few minutes) and proposed that degradation continued via the dechlorination and dealkylation pathways. The disappearance of carbaryl under phosphate buffer in deionized nondeoxygenated water (pH 6.6) was also observed by Ghauch et al. (2001). In an earlier study, Sayles et al. (1997) demonstrated the dechlorination of the highly recalcitrant pesticides DDT, DDD, and DDE by using ZVI under anaerobic conditions at pH_0 of 7.

ZVI may be used to treat higher contaminant loads that are resistant to biodegradation (Bell et al., 2003), as the technology is not susceptible to inhibition that microorganism sometimes encounter with chlorinated compounds. Even poly-halogenated pollutants can be destroyed via reductive dehalogenation using ZVI in contrast to many advanced oxidation processes (AOPs) such as H_2O_2+UV, Fenton, photolysis, O_3, O_3+UV (Pera-Titus et al., 2004), UV, UV/H_2O_2, and Photo-Fenton (Al Momani et al., 2004). The presence of oxygen is generally assumed to lower the efficiency of the reduction process as a result of competition with the target contaminants, e.g. organics or metals with the reduction of oxygen by ZVI generally envisaged as a four-electron step with water as the major product:

$$2Fe^0 + O_2 + 2H_2O \rightarrow 2Fe^{2+} + 4OH^- \tag{2.3}$$

Additionally, further oxidation of Fe^{2+} to Fe(III) species is likely with subsequent precipitation of particulate iron oxyhydroxides, which may coat the Fe^0 surface and lower the reaction rate. Consistent with such an effect, Tratnyek et al. (1995) found

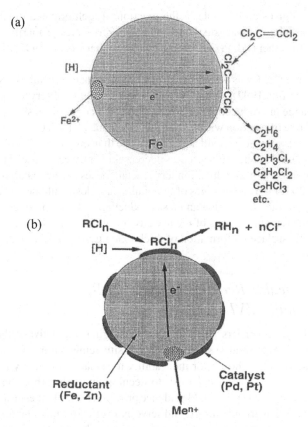

FIGURE 2.1. (a) Perchloroethylene dechlorination (Zhang et al., 1998). (b) A nanoscale bimetallic particle for chlorinated solvent removal (Zhang et al., 1998).

that the half-time for dechlorination of 180 μM carbon tetrachloride by 16.7 g/L of 325 mesh Fe^0 granules increased from 3.5 h when reaction mixtures were purged with nitrogen to 111 h when purged with oxygen. Surprisingly, Tratnyek et al. (1995) observed a higher rate of degradation of CCl_4 in an air-purged system ($\tau_{1/2} = 48$ min) than in the nitrogen-purged case ($\tau_{1/2} = 3.5$ h). It would thus appear that the impact of oxygen in ZVI-mediated degradation of organic compounds is worthy of further investigation.

2.1.2. Nanoparticulate Bimetallic and Iron Technology

In addition to transformation by Fe^0, bimetallic coupling with a second catalytic metal has also been used in degrading a variety of contaminants as environmental cleanup. In most cases, rates of transformation by bimetallic combinations have been significantly faster than those observed for iron metal alone (Appleton, 1996; Fennelly and Roberts, 1998; Muftikian et al., 1996; Wan et al., 1999).

Figure 2.1b gives an example of the reaction of a chlorinated organic with a bimetallic particle. In this system one metal (Fe, Zn) serves primarily as electron donor while the other (Pd, Pt) serves as a catalyst (Cheng et al., 1997; Zhang et al., 1998).

Of many metals, Cu is known as a mild hydrogenation catalyst (Satterfield, 1991; Yang *et al.*, 1997). Fennelly and Roberts (1998) observed that a more dramatic change in product distribution is seen in the copper/iron system than that in increased rate of reaction with 1,1,1-trichloroethane (1,1,1-TCA) by nickel/iron and the bimetals showed a significantly faster rate than only iron. The effectiveness of the catalyst used in bimetallic process decreases over time because of the buildup of an iron hydroxide film, which hinders reactant access to the catalytic sites (Li and Farrell, 2000). The advantages of bimetallic particles would be higher activity and stability for the degradation and less production of toxic intermediates; however, concerns remain in terms of the toxicity in catalytic metals and deactivation of the catalytic surface by formation of thick oxide films (Muftikian *et al.*, 1996).

2.1.3. Permeable Reactive Barrier (PRB) Using Granular ZVI

Permeable reactive barriers (PRBs) are an emerging alternative technology to traditional pump-and-treat systems for the in situ remediation of groundwater. Reactive materials are chosen for their ability to remain sufficiently reactive for periods of years to decades and work to dechlorinate halocarbons via reaction (2.2) (Benner et al., 1997). The field evidence provided by O'Hannesin and Gillham (1998) indicates that granular iron could serve as an effective medium for the in situ treatment of chlorinated organic compounds in groundwater. The iron was placed in the ground as a PRB (Figure 2.2); other configurations place the iron within the reaction cells of funnel-and-gate systems, around the exterior of a pumped well, or at treatment points in an impermeable encasement around hazardous waste (Reeter, 1997).

The most common methods of installation include constructing a trench across the contaminated groundwater flow path by using either a funnel-and-gate system or a continuous reactive barrier (NFESC, 2004). The gate or reactive cell portion is typically filled with granular ZVI. There are several methods for emplacing PRBs, including trench and fill, injection, or grouting (USDEGJO, 1989).

There are many advantages of using passive reactive barriers compared with existing ex situ treatment technologies. PRBs require no external energy source, and it is possible that iron fillings may last 10–20 years before requiring maintenance or replacement (NFESC, 2004). Studies have shown that iron barriers are more cost effective than pump-and-treat systems (Day et al., 1999; Fruchter et al., 2000; USDEGJO, 1989). For instance, although the installation of a PRB requires a higher initial capital investment, operating and maintenance (O&M) costs are significantly lower, provided that the PRB does not show an unexpected breakdown

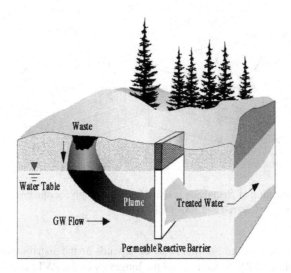

FIGURE 2.2. Typical configuration of a permeable reactive barrier (PRB), showing the source zone, plume of contamination, treatment zone, and plume of treated groundwater (Powell et al., 1998).

before costs are recovered (Birke et al., 2003; Powell et al., 2002). While the advantages of ZVI barriers are compelling, the long-term problems are not well understood and may include chemical and/or biological precipitate formation at the barrier, changes in contaminant removal efficiency over time, consumption of dissolved oxygen, higher pH, and modification to the groundwater hydraulic conductivity (Powell and Puls, 1997; Puls et al., 1999; USDEGJO, 1989). The lifetime of PRBs using Fe^0 as a reactive medium is expected to be primarily limited by precipitation at the barrier (Liang et al., 2003). There are also concerns regarding the maintenance, lifetime, and costs of this technology (Felsot et al., 2003). Nevertheless, PRBs have the potential to gain broad acceptance (Birke et al., 2003).

2.1.4. PRB and ZVI Colloids

Another approach to the installation of passive reactive barriers involves injection of ZVI colloids into porous media (e.g., the subsurface environment). In such systems, colloidal barriers are placed in the subsurface environment, perpendicular to groundwater flow, and selectively remove targeted groundwater contaminants as water whereas other nontargeted constituents pass through the barrier as shown in the Figure 2.3 (Kaplan et al., 1996). As illustrated in the figure, the movement of colloidal ZVI can be controlled to some extent by injecting the colloids in one well and withdrawing groundwater from a nearby second well, thereby drawing colloids in the desired direction.

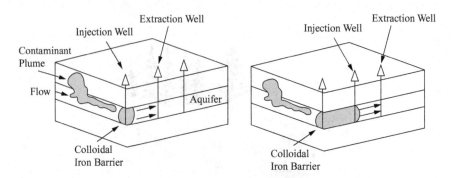

FIGURE 2.3. Formation of chemically reactive barrier through coordinated use of injection and extraction wells (Cantrell and Kaplan, 1997).

The effectiveness of the chemical barrier depends on the distribution and uniformity of the colloidal ZVI injection and the longevity of the ZVI materials. Cantrell and Kaplan (1997) observed in column experiments that as the injection rate was increased, the Fe^0 concentration became more uniform. They predicted that the life span of the barrier would be 32 years based on groundwater flow rate, effective porosity, and barrier thickness. The reported advantages of colloidal barriers are that there are no requirements for above-ground treatment facilities, installation is relatively simple, capital costs are moderate, and there are no additional waste disposal requirements. However, for long-term performance, the total mass of reactive material, rate of reaction within the barrier, and physical changes such as decreases in porosity and permeability may limit the lifetime of the barrier. Performance will also depend on the nature of contaminants, groundwater flux, subsurface geology, and chemistry (Kamolpornwijit et al., 2003).

2.1.5. Use of ZVI, H_2O_2, and Complexants

The Fenton reaction consumes H_2O_2 in the following redox reaction giving rise to the potent hydroxyl radical:

$$Fe^{II} + H_2O_2 \rightarrow Fe^{III} + HO^- + HO^\bullet \qquad (2.4)$$

The hydroxyl radical reacts with almost any organic species with diffusion-limited kinetics. Production of Fe^{II} complexes occurs through the corrosion of Fe(0) via reaction (2.1).

There are several reports of using a combination of ZVI and peroxide/complexants to promote remediation of water and soil highly contaminated with organics. Hundal et al. (1997) showed that ZVI combined with H_2O_2 destroyed 2,4,6-trinitrotoluene (TNT) and hexahydro-1,3,5-trinitro-1,3,5-triazine (RDX) in contaminated soil slurries more efficiently than ZVI alone. Less iron was required to achieve the same level of remediation. For example, sequential treatment of a TNT-contaminated solution (70 mg TNT/L spiked with ^{14}C-TNT) with ZVI (5%

w/v) followed by H_2O_2 (1% v/v) completely destroyed TNT and removed about 94% of the ^{14}C from the solution, 48% of which was mineralized to $^{14}CO_2$ within 8 h. It was shown that adding ethylenediaminetetraacetic acid (EDTA) or glucose to a Fe(0)-amended TNT solution resulted in 6% mineralization, while only 0.5% mineralization of untreated TNT was observed. Similar improvements were observed by Noradoun et al. (2003), who demonstrated the complete destruction of 4-chlorophenol and pentachlorophenol in the presence of Fe(0), EDTA, and O_2 (aq) in the absence of added H_2O_2.

The feasibility of Fenton's oxidation of methyl *tert*-butyl ether (MTBE) using ZVI as the source of catalytic ferrous iron was assessed in a study by Bergendahl and Thies (2004). More than 99% of MTBE-contaminated water was removed at pHs 4 and 7 using a H_2O_2/MTBE molar ratio of 220:1. Similarly, Lücking et al. (1998) investigated the oxidation of 4-chlorophenol in aqueous solution by hydrogen peroxide in the presence of a variety of additional substrates including iron powder. H_2O_2 oxidation of 4-chlorophenol in the presence of iron powder proceeded much faster when iron powder was used instead of graphite or activated carbon, presumably via Fenton's oxidation of the 4-chlorophenol. Studies by Tang and Chen (1996) showed the degradation of azo dyes was faster using the H_2O_2/iron powder system than the Fenton's reagent system, e.g., H_2O_2/Fe(II). The difference was attributed to the continuous dissolution of Fe(II) from the iron powder and the dye adsorption on the powder.

Another system using peroxide involves the combination of hydrogen peroxide and electrochemically amended iron, which has been found to successfully degrade the two organophosphorous insecticides malathion and methyl parathion (Roe and Lemley, 1997). In this system, Fe(II) is generated electrochemically at the Fe(0) electrode while H_2O_2 can be either added from an external source or generated by reduction of oxygen at mercury or graphite. It appears that the addition of H_2O_2 in these studies initiates the Fenton reaction and results in oxidation of organic contaminants. Hundal et al. (1997) note that the Fe(0)-treated contaminants could be more susceptible to biological mineralization than would otherwise be the case.

2.1.6. Nanosized ZVI (nZVI)

Zhang et al. (1998) at Lehigh University investigated the application of nanosized (1–100 nm) ZVI particles for the removal of organic contaminants and found that not only is the reactivity higher due to an elevated surface area (average of 33.5 m^2/g for the nanosized particles compared with 0.9 m^2/g for the commonly used microscaled particles) but the reaction rate is also significantly higher (by up to 100 times) on a surface area normalized basis. In one such system, 1.7 kg of nZVI particles were fed into a 14-m^3 groundwater plume over a 2-day period, as illustrated in Figure 2.4 (Elliot and Zhang, 2001). Despite the low particle dosage, trichloroethylene reduction efficiencies of up to 96% were observed over a 4-week monitoring period, with the highest values observed at the injection well and at adjacent piezometers in the well field. The critical factors that influence

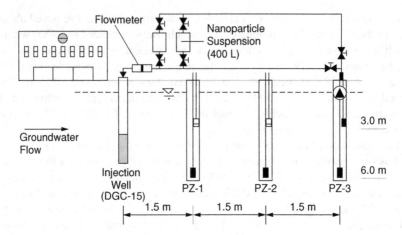

FIGURE 2.4. Schematic of in situ injection of nanoscale bimetallic particles (Elliott and Zhang, 2001).

degradation kinetics appear to be the ZVI condition and available surface area (Chen et al., 2001; Choe et al., 2000).

Nanoparticles may provide an effective, flexible, and portable remedial technique for suitable groundwater contaminants such as chlorinated hydrocarbons (Elliot and Zhang, 2001). Since the reactions with organohalides are often thought to be "inner-sphere" surface-mediated processes, the use of nanometer-sized iron particles is therefore a real potential advantage.

Other possibilities include remediation of on-farm irrigation channels or dams (for pesticide contamination) or remediation of contaminated sites where surface application with subsequent infiltration would appear feasible (Feitz et al., 2002). The particles could also be attached to activated carbon, zeolite, or silica, with the added advantage of the adsorptive removal of polycyclic aromatic hydrocarbons (PAHs) and other highly persistent contaminants such as chlorinated hydrocarbons (CHCs) (Birke et al., 2003).

2.2. Pesticides and Contamination

2.2.1. Introduction

Pesticides and herbicides are used extensively in agricultural production throughout the world to protect plants against pests, fungi, and weeds. For example, world pesticide usage exceeded £5.6 billion and expenditures totaled more than US$33.5 billion in 1998–1999 (Donaldson et al., 2002). Pesticide usage during grain production is particularly high, and in the grain belt of southwest Western Australia, central and southern Queensland, and northeast New South Wales, total expenditure on crop chemicals was estimated at more than $50,000 in 1998–1999. In contrast, the expenditure in the rest of the mixed farming regions ranges from $5,000 to $30,000 (Australia State of the Environment Committee, 2001).

The Indian pesticides industry is the largest in Asia and twelfth largest in the world with a value of US$0.6 billion, which is 1.6% of that of the global market (IPISMIS, 2001; Mindbranch, 2001). The continuous growth of the pesticide industry in India has contributed to the worsening problems of air, water, and soil pollution in this country (Mall et al., 2003). China is also a large producer and consumer of pesticides (Qiu et al., 2004). From 1949 on, the consumption of pesticides in China increased rapidly from 1920 ton in 1952 to 537,000 ton in 1980, and then decreased to 271,000 ton in 1989 after the manufacture of organic chlorinated pesticides ceased at the beginning of the 1980s (Li and Zhang, 1999).

In Australia, cotton production has been particularly successful and is currently worth approximately $1.5 billion per year (Raupach et al., 2001). The substantial growth in the cotton industry, however, has resulted in environmental contamination. For example, in the irrigated cotton region of central and northern New South Wales, the presence of several pesticides has been detected in rivers near and downstream of cotton-growing areas during the growing season (Raupach et al., 2001). In particular, spot-sampled riverine concentrations of the insecticide endosulfan were found to range from 0.02 to 0.2 ppb, which significantly exceeds environmental guidelines for protection of ecosystems (currently 0.01 ppb; Australian and New Zealand Environment and Conservation Council, 1992).

The extensive use of pesticides affects the wider ecology and there are links with birth defects in birds and fish (Ferrano et al., 1991; McKim, 1994; Nowell et al., 1999; Oliver, 1985). High deposition of pesticides in a sediment can inhibit the microbial activity in the sediment (Redshaw, 1995), and certain pesticides such as α-BHC, γ-BHC, isodrine, dieldrin, and p-p'-DDT accumulate in fish (Amaraneni, 2002). Pesticides also have cumulative effects on the human body and lead to several diseases, ranging from chronic common cough and cold to bronchitis and cancer of the skin, eye, kidney, and prostate gland (Gupta and Salunkhe, 1985; Paldy et al., 1988).

2.2.2. Characteristics of Pesticides and Their Environmental Effects

The recalcitrance of a pesticide is largely determined by its chemical structure. Some herbicides (such as 2,4-D) are susceptible to environmental degradation, while others (including most chlorinated insecticides such as endosulfan, heptachlor, and dieldrin) are considerably more resistant. Solubility will affect not only transport but also pesticide degradation since degradation is believed to occur mainly in the solution phase. The characteristics and structures of pesticides investigated in this research are presented in Table 2.1 and Figure 2.5. Ionizability, water solubility, volatility, soil retention, and longevity are key properties.

TABLE 2.1. Chemical and physical properties of commonly used pesticides and banned pesticides still routinely found in contaminated soils (Hartley and Kidd, 1983)

Compound	Formula (molecular weight)	$T_{1/2}$ (days)	Solubility in water (mg/L)	Log octanol/ water partition coefficient	Vapor pressure (mbar)
Atrazine	$C_8H_{14}ClN_5$ (216)	60^a	33	2.7	4×10^{-7} (20°C)
Aldrin	$C_{12}H_8Cl_6$ (365)	$>30, <100^b$	<0.05	5.11	3×10^{-5} (20°C)
Dieldrin	$C_{12}H_8Cl_6O$ (381)	$1460-2555^c$	<0.1	3.7–6.2	10^{-6} (20°C)
Diuron	$C_9H_{10}Cl_2N_2O$ (233)	90^a	42	2.8	4×10^{-6} (50°C)
Molinate	$C_9H_{17}NOS$ (187)	21^a	880	2.9	7×10^{-3} (25°C)
Chlorpyrifos	$C_8H_{11}Cl_3NO_3PS$ (351)	30^a	2	5.0	2×10^{-5} (25°C)
Heptachlor	$C_{10}H_5Cl_7$ (373)	$>30, <100^b$	0.056	5.44	4×10^{-4} (25°C)
Chlordane	$C_{10}H_6Cl_8$ (410)	$>100^b$	0.056	2.78	1×10^{-5} (25°C)
Diazinon	$C_{12}H_{21}N_2O_3PS$ (304)	40^a	40	3.3	2×10^{-4} (20°C)
Endosulfan	$C_9H_6Cl_6O_3S$ (407)	50^a	0.32	N/A	1×10^{-2} (80°C)

[a] Adapted from Weber (1994)
[b] PMEP (2003)
[c] WHO (1989)

Pesticides are widely distributed in drinking waters, groundwaters, and soils. There are various routes for pesticide contamination in the environment including runoff from agricultural land, direct entry from spray, industrial effluents, and dust. Pesticide contamination of soils, water, and other matrices may also be caused by accidental spills during manufacture, formulation, and shipment or at local agrochemical dealerships. Although many current pesticides are designed to break down quickly in sunlight or in soil, they are more likely to persist if they reach groundwater because of reduced microbial activity, absence of light, and lower temperatures in the subsurface zone (National Center for Toxic and Persistent Substances, 1995).

Residues of pesticides have significant environmental impacts on aquatic ecosystems and mammals. For example, in the drainage and irrigation canals in southern New South Wales, Australia, high concentrations of pesticides (e.g., molinate) have been regularly detected (Australia State of the Environment Committee, 2001). Such pesticides (particularly endosulfan) have been linked to large fish kills in several rivers throughout Australia. Freshwater crustaceans are particularly at risk (Australia State of the Environment Committee, 2001). Besides the detrimental effect on natural ecosystems, there are negative economic and social impacts associated with agrochemical contamination of both irrigation networks and the wider aquatic environment. There are instances where it is simply not possible to hold water or where uncontrolled releases result in a severe reduction in the quality of irrigation waters as shown in the Table 2.2. A major economic concern of elevated pesticide levels in irrigation channels is that the water may contain pesticides that are incompatible and harm crops for users downstream of uncontrolled releases. For example, atrazine used by citrus and sorghum growers is toxic to soybeans.

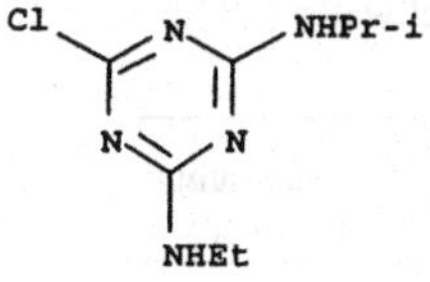

Atrazine

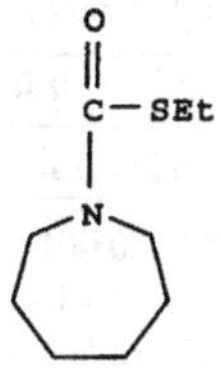

Molinate

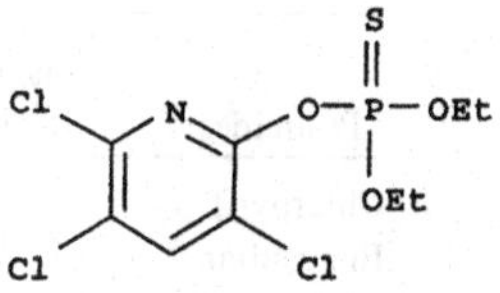

Chlorpyrifos

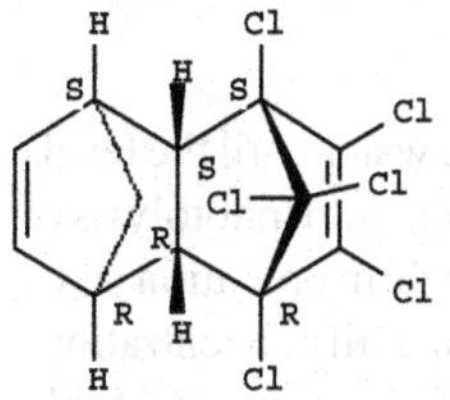

Aldrin

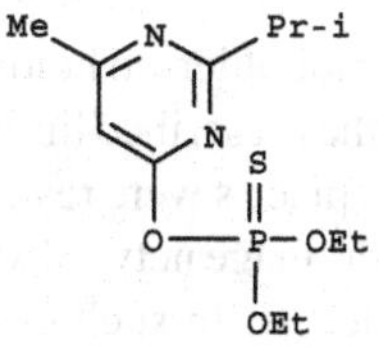

Diazinon

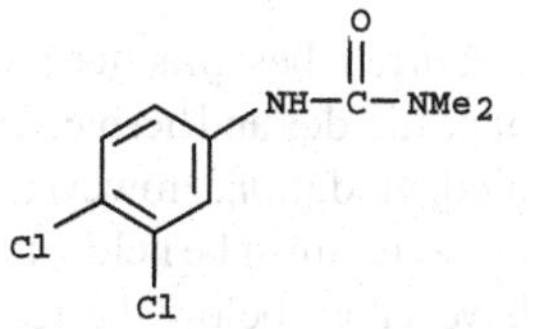

Diuron

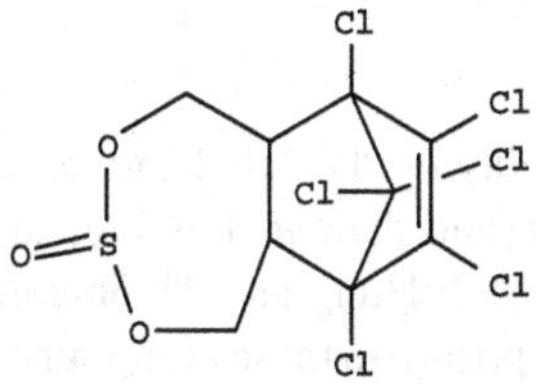

Endosulfan

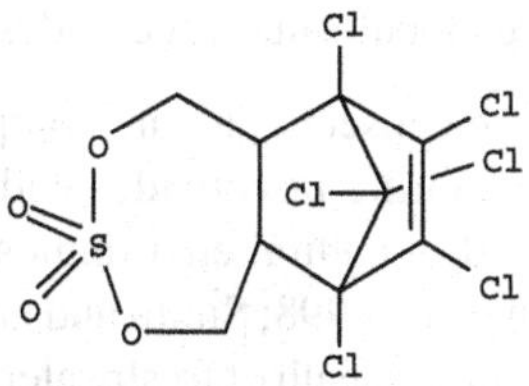

Endosulfan sulfate

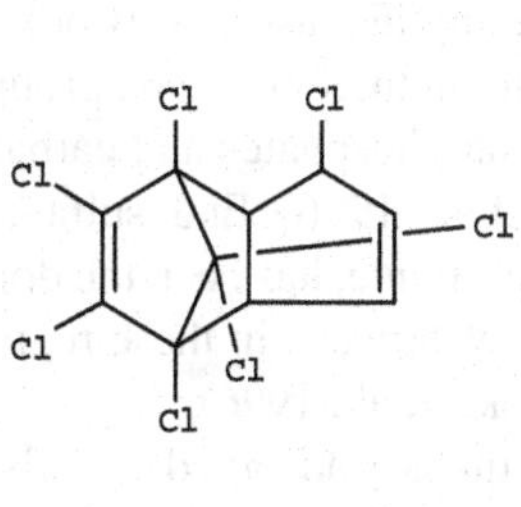

Heptachlor

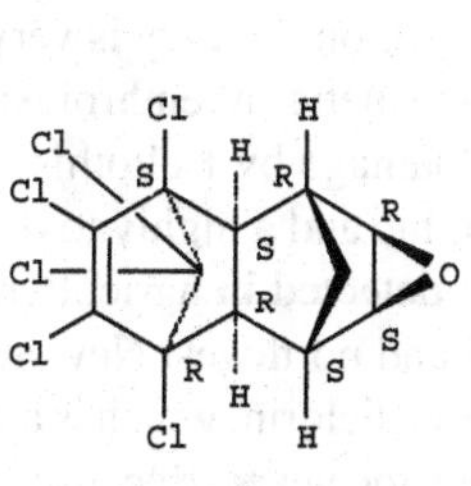

Dieldrin

FIGURE 2.5. Chemical structures of compounds investigated in this research.

TABLE 2.2. Example of highly contaminated irrigation channel water
(Data courtesy of CSIRO Land and Water, 2001)

Pesticides	Concentration found in irrigation channel (μg/L)	EPA Limit (μg/L)		Toxicity ratings (Q value)
		Notification	Action	
Chlorpyrifos	25	0.001	0.005	9,520
Endosulfan	0.58	0.05	0.1	4,700
Molinate	30	12.5	25	12
Malathion	30	0.07	0.35	340
Diuron	700	8	40	9
Atrazine	79	2	10	3

Current best practice requires landholders to hold discharge water until the levels of pesticides and herbicides meet the prescribed limits through natural photolysis or biodegradation. However, this can place severe restrictions on farm operation since the water must be held within the drainage network on the farm until concentrations have fallen below the regulated levels. In such cases, an inexpensive but highly effective treatment technology that could remove the pesticides before they are released and thus prevent detrimental downstream effects would be useful.

2.2.3. Commonly Used Pesticides

2.2.3.1. Organochlorine Insecticides

Organochlorine insecticides are compounds that are highly lipid soluble and toxic. The organochlorine insecticides endosulfan and aldrin and their metabolites are often detected in natural environment (Guerin et al., 1992; Hung and Thiemann, 2002; Matin et al., 1998; Smith and Gangolli, 2002). Exposure to these compounds has resulted in the death of freshwater species (Mishra and Shukla, 1997; Naqvi and Vaishnavi, 1993) and bioaccumulation in organisms, which may produce adverse effects on ecosystems (Hutson and Roberts, 1990). Of the organochlorine insecticides, endosulfan is in the most widely used in Australia, the United States, and elsewhere and has been used widely in cotton farming in Australia (Brooks et al., 1996). The cotton industry is very much dependent on three pesticide groups (endosulfan; synthetic pyrenthroids; and certain organophosphates and carbamates) to prevent damage by Heliothis species (Brooks et al., 1996). Endosulfan, which is hydrophobic and a highly toxic and hazardous pesticide, has been the dominant insecticide detected in agricultural areas (natural waterways in these regions) of the central and northwest New South Wales (Brooks et al., 1996).

Aldrin and dieldrin, which is aldrin epoxide, are quickly adsorbed on soils where they remain for years. Because of their high persistence and toxicity, the use of aldrin and dieldrin, which were used in soil treatment after harvest (grape vines, bananas) (INCHEM, 1989) and for termite controls (Stevenson et al., 1999), was banned in 1995. They are still detected in the environment, however, because of their persistence and previous wide use as insecticides for the control of pests on crops such as corn and cotton. Aldrin and dieldrin are structurally similar synthetic compounds, highly toxic and hazardous for humans (e.g., toxic by mouth, skin contact, inhalation of dust) and aquatic and terrestrial life. Aldrin is readily converted to dieldrin under normal environmental conditions (Ramamoorthy, 1997) and, as a result, dieldrin residues in soil are higher than those of aldrin. Dieldrin is one of the most persistent of the chlorinated hydrocarbons and is highly resistant to biodegradation (UNEP, 1989). Abiotic processes play a limited role in the degradation of aldrin and dieldrin in the environment (WHO, 1989).

Heptachlor and chlordane, the use of which has now been banned, were used in termite and ant control (ATSDR, 1992, 1993), as well as pest control on cotton crops (ATSDR, 1990). Heptachlor was used primarily as an insecticide in seed grains and on crops during the 1960s and 1970s before it was banned in 1995. The microbial and photochemical transformation products heptachlor epoxide and photoheptachlor remain in soil for long periods of time (>15 years) and are equally or more toxic than the parent compound (Ramamoorthy, 1997). Heptachlor is fairly stable to light and moisture and it is not readily dehydrochlorinated. Its half-life in the soil in temperate regions ranges between $3/4$ and 2 years. It is not likely to penetrate into groundwater, but contamination of surface water and sludge can occur (WHO, 1989). Chlordane is another toxic organochlorine pesticide that was used routinely from 1948 to 1988. Chlordane is not a single compound, but a mixture of about 10 major compounds (Ramamoorthy, 1997), and is highly persistent in the environment (see Table 2.1).

2.2.3.2. Herbicides

Atrazine is one of the most widely used herbicides in the United States, Europe, and Australia and is still used for control of annual broadleaf weeds and certain annual grasses, particularly in corn production (KDARFC, 2004). Atrazine is the most commonly detected pesticide in the river systems in Australia (Harris and Kennedy, 1996). It is moderately soluble and, because of its persistence in water and mobility in soil, is among the most frequently detected pesticide in groundwater (Ghauch and Suptil, 2000).

Diuron is a widely used herbicide, because of its ability to inhibit photosynthesis (Mazellier et al., 1997). It is also used for control of a wide variety of annual and perennial broadleaf and grassy weeds and is used on many agricultural crops such as fruit, cotton, sugarcane, alfalfa, and wheat (Goody et al., 2002; Macounová et al., 2003; Råberg et al., 2003). It is stable in neutral media at normal temperatures but is hydrolyzed by acids and alkalis and at elevated temperatures (diuron decomposes at 180–190°C). The degradation of diuron through chemical (hydrolysis) or biological processes is very slow at neutral pH. Because of its chemical stability and moderate solubility, diuron is often detected in surface waters and groundwaters (Mazellier and Sulzberger, 2001).

Molinate is one of five thiocarbamate herbicides, a class of compounds that possess low volatility and are slowly degraded by hydrolysis over a period of months (WHO, 1988). It is a moderately toxic herbicide used extensively worldwide in the rice industry for the control of germinating broad-leaved and grass weeds, particularly *Echinochloa* spp. (Hsieh et al., 1998). Molinate is only weakly bound to soils, is soluble in water and mobile, and presents a significant contamination risk to groundwaters.

2.2.3.3. Organophosphorus Insecticides

Chlorpyrifos is an organophosphorus pesticide that is widely used in the home to control cockroaches, fleas, and termites and in some pet flea and tick collars. Chlorpyrifos is also used on grain, cotton, field, nut, and vegetable crops (Cochran et al., 1995) and on the farm, as a dip or spray to control ticks on cattle, and as dust or spray to control pests on crops such as rice, fruit, vineyards, sugarcane, corn, tobacco, potatoes, and other horticultural crops (Ramamoorthy, 1997; WHO, 1998). Typical field dissipation half-life at the soil surface is 1–2 weeks and for soil-incorporated applications (when applied to high organic matter soil), 4–8 weeks. The half-life of chlorpyrifos in water is relatively short, from a few days to 2 weeks (US EPA, 2000).

Diazinon is another organophosphorus insecticide with a wide range of insecticidal activity. Diazinon has been widely used with applications in agriculture and horticulture for controlling insects in crops, lawns, fruit, and vegetables and as a pesticide in domestic, agricultural, and public buildings (NRA, 2000; Worthing and Hance, 1991). It is stable in neutral media but slowly hydrolyzes in alkaline media and more rapidly in acid media. In natural water, diazinon has a half-life of the order of 5–15 days (WHO, 1998).

2.2.4. Pesticides Treatment and Management Practices

Pesticides play a critical role in worldwide agriculture, but uncontrolled releases are a major environmental concern. Remediation of soil and water contaminated with pesticides range from conventional treatment techniques (e.g. incineration, thermal desorption, soil flushing/washing, bioremediation, land-farming, phytoremediation, photochemical processes, and direct oxidative processes) to innovative remediation technologies such as ultrasound-promoted remediation and other advanced oxidation technologies. The properties of pesticides in soil and water strongly influences disposal options, as does treatment costs, public health, and technical feasibility. The major techniques for the remediation of contaminated soils, surface waters, and groundwaters are described in more detail below.

Incineration and thermal desorption: Treatment by incineration reduces the volume and destroys toxic materials in contaminated soils that may otherwise remain for hundreds of years. Incineration affects treatment of a contaminated soil through two mechanisms: desorption, which removes the pesticide from the soil and liberates it to the gas phase, and combustion, which destroys the target compound (Stevenson, 1998). Most pesticides are thermally fragile and therefore amenable to incineration. Thermal desorption is the most widely used treatment for cleaning up contaminated soil from large-scale sites (Troxler, 1998). Pesticide removal efficiencies from soil are greater than 99% for most pesticides using a typical thermal desorption system (Troxler, 1998). Both techniques, however, are expensive options for soil remediation and generally lack public acceptance because of the health concerns of nearby residents.

Soil flushing and washing: Soil flushing and washing are processes that employ water, cosolvents, surfactants, or supercritical fluids to remove organic contaminants from soils. Supercritical fluid extraction (SFE) using carbon dioxide alone, or in combination with a modifier, has been shown to be an effective extraction method for pesticides in several matrices although the cost was estimated to be about twice the cost of incineration (Rock et al., 1998). Pesticides commonly degraded by SFE are 2,4-D (Rochette et al., 1993) and organochlorine pesticides (Barnabas et al., 1994; Nerín et al., 2002).

Phytoremediation and bioremediation: Phytoremediation, which uses plants to clean up contaminated environments such as soil, water, or sediments, is potentially more cost effective and less environmentally disruptive than conventional ex situ remediation technologies (Schnoor et al., 1995; Wenzel et al., 1999). Nevertheless, recalcitrant halogenated organic chemicals such as DDT, dieldrin, and PCBs are bound tightly to the soil and have low water solubility, resulting in very little of the residue being taken up into plants. Bioremediation is an uncontrolled process that can be stimulated with selective nutrients or fortified by bioaugmentation and involves inoculating sites lacking the appropriate strain(s) with nonindigenous pesticide-degrading microorganisms (Van Veen et al., 1997).

The feasibility of bioremediation depends on the specific contaminant and its suitability as a substrate for microbial degradation. The planned future use of the site is also an important consideration (Arthur and Coats, 1998). Detailed site characterization and preliminary feasibility studies are required for the design and optimization of any biostimulation approach. Remediation also depends on the site-specific nature of each contaminated matrix (Zablotowicz et al., 1998). A bioactive soil barrier technique, known as the Filter technique, which combines the use of contaminated water with filtration through the soil to a subsurface drainage, has been found to reduce pesticide loads by up to 99% (Jayawardane et al., 2001). However, field studies have shown that the concentration of pesticides in the discharge, particularly mobile ones such as molinate, are often found above accepted environmental limits (Biswas et al., 2000).

Land-farming: Land farming involves mixing or dispersing wastes into the upper zone of the soil-plant system with the objective of microbial stabilization, adsorption, immobilization, selective dispersion, or crop recovery. Land farming is an older, proven bioremediation technology that can be applied to pesticide waste. Land farming is commercially applied for the remediation of pesticide waste at agrochemical retail facilities under special state permits (Andrews Environmental Engineering, Inc., 1994).

2.2.4.1. Advanced Oxidation Processes

Advanced oxidation processes (AOPs) appear to be suited to the treatment of pesticide-containing waste. Indeed, many hundreds of laboratory studies have shown that pesticides may be oxidized using AOPs such as UV/H_2O_2, photocatalysis, ozonation, and Fenton-based processes. The efficiency of the oxidation process, however, is a function of the type and nature of the waste and the structural properties of the pesticides (Larson and Weber, 1994).

AOPs are based on hydroxyl radical generation and subsequent oxidation of the organic substrate. The efficiency of AOPs strongly depends on operating parameters such as pH, initial pesticide concentration, solubility, and light intensity for photochemical processes. For example, the degradation rate of a pollutant in the UV/H_2O_2 system is affected by the H_2O_2 concentration, UV light intensity, and, to a lesser extent, solution pH. As one of modifications of Fenton's reagent reaction (2.4), Sun and Pignatello (1992) observed complete mineralization of phenoxyacetic herbicides in a Fe^{3+}/H_2O_2 system, where reaction rate is dependent on the concentrations of H_2O_2 and chelators and the pH. However, in practice, oxidation has been limited to waste containing low organic matter because other constituents can act as radical scavengers and lower the effectiveness toward trace contaminant degradation.

Photochemical AOPs often induce rapid degradation through homogeneous (e.g., UV/H_2O_2) (Muszkat, 1998), Fe^{3+}/UV, or heterogeneous processes (e.g., TiO_2/UV, ZnO/UV) (Legrini et al., 1993). Although photochemical AOPs have definite advantages over other AOP methods, further development of more active, less costly photocatalysts and increase in the efficiency of sunlight/UV lamp utilization are required before widespread adoption of the technology becomes likely.

Ozonation processes and chemical oxidation processes appear to be especially suitable for industrial applications. The degradation products formed during AOPs of hydrophobic pesticides are often more polar and more bioavailable than the parent compounds. Complete mineralization can often be enhanced by coupling AOPs to biodegradation (Chiron et al., 2000).

High-power ultrasound can promote both oxidation and reduction through the formation of OH radicals (powerful oxidizing agents) and H radicals (effective reducing agents) during the thermal dissociation of water ($H_2O \rightarrow H^{\bullet} + OH^{\bullet}$) (Yak et al., 1999). This process is one of the most intriguing and least obvious advanced treatment methods for chemical wastes. However, practical application of sonochemistry to chemical waste treatment has proven to be challenging and limited its application. For example, the energy efficiency and economics of sonochemical treatment need to be better defined, and practical production-scale reactors need to be developed (Sivakumar and Gedanken, 2004). The design of reactors and maintenance of treatment efficiency under practical conditions are likely to be difficult. Also, the fundamental physical and chemical processes occurring in sonochemical treatment remain less well defined than for most other advanced treatment processes (Rock et al., 1998).

2.2.4.2. Zero-Valent Metal Remediation

Initial screening studies using nanoscale ZVI particles found that cyclodiene insecticides (e.g., chlordane, heptachlor, aldrin, dieldrin, endosulfan sulfate, α-, β-endosulfan) are generally very resistant to degradation by ZVI (Table 2.3) (Waite et al., 2004). The hydrophobic nature of organic pollutants, particularly

TABLE 2.3. Summary of the extent of degradation for pesticides by ZVI.

Compound	Chemical formula	Condition	Initial concentration (μg/L)	Total reaction time (h)	Fe^0 (mM)	pH	Analysis technique	Percent removal
Atrazine	$C_8H_{14}ClN_5$	Anaerobic	1000	7	36	6.8[a]	GC/MS	84
Aldrin	$C_{12}H_8Cl_6$	Aerobic	20 (10)	6 (29)	5.4 (8.9)	6.8	ELISA	50 (70)
Dieldrin	$C_{12}H_8Cl_6O$	Aerobic	10	5	8.9	6.8	ELISA	0
Diuron	$C_9H_{10}Cl_2N_2O$	Aerobic	100	5	18	6.8	ELISA	12
Molinate	$C_9H_{17}NOS$	Aerobic	100	3	11	NB[b]	GC/MS	>99
Chlorpyrifos	$C_8H_{11}Cl_3NO_3PS$	Aerobic	10	0.25	25	NB	ELISA	98
Heptachlor	$C_{10}H_5Cl_7$	Aerobic	10	5	8.9	6.8	ELISA	0
Chlordane	$C_{10}H_6Cl_8$	Aerobic	10	5	8.9	6.8	ELISA	0
Diazinon	$C_{12}H_{21}N_2O_3PS$	Aerobic	0.1	3	11	NB	ELISA	27
Endosulfan sulfate	$C_9H_6Cl_6O_4S$	Aerobic	10	27	8.9	6.8	ELISA	25
α-endosulfan	$C_9H_6Cl_6O_3S$	Aerobic	10	5	8.9	6.8	ELISA	0
β-endosulfan	$C_9H_6Cl_6O_3S$	Aerobic	10	5	8.9	6.8	ELISA	0

[a] pH 6.8 set using a 0.02 M phosphate buffer
[b] pH not buffered in these cases because of interference (Waite et al., 2004)

halogenated organic compounds, seems to limit the efficient electron transfer due to their immiscibility with water.

While ZVI was not effective in degrading endosulfan, it did prove effective for other pesticide and herbicides. These include compounds containing nitrogen heteroatoms such as atrazine, molinate, chlorpyrifos, and, to a limited extent, diazinon and diuron (Table 2.3). Chemicals with seemingly similar structures reacted differently. This is especially true for aldrin and dieldrin, where the chemical structure is quite similar but the extent of degradation is significantly different. A similar, although less pronounced, effect was observed for α- and β-endosulfan compared to endosulfan sulfate.

Agrawal and Tratnyek (1996) observed that nitro-substituted entities were reduced by ZVI significantly faster than by dechlorination. Some researchers have noted that the treatment of strongly surface-active organic chemicals such as PCBs, dioxin, DDT, toxaphene, mirex, lindane, and hexachlorobenzene may not be practicable using ZVI (Weber, 1996). The results of the screening study undertaken here suggest that molinate is particularly well suited to ZVI-mediated degradation. Molinate is an indicator chemical used by the New South Wales Environment Protection Authority (EPA) for flagging likely pesticide contamination in irrigation channels. Molinate is highly soluble and has an environmental half-life of 21 days (Hartley and Kidd, 1983). It is one of the most heavily used herbicides in Australia and is routinely detected in waterways in Australia (Australia State of the Environment Committee, 2001). Because molinate undergoes slow hydrolysis, it may also leach into and persist in groundwaters. Additional details of the screening studies in degradation of different pesticides herbicides using nanoscale ZVI are given in Appendix E.

2.3. Summary

Fe^0 has been reported to be very effective for the reduction of various organic and inorganic contaminants. Granular ZVI incorporated into permeable reactive barriers (PRBs) has proven to be a cost-effective in situ remediation method for groundwaters contaminated with chlorinated organics and appears to be a particularly promising alternative technology to traditional pump-and-treat systems. Nanosized ZVI has a greater reactivity than granular ZVI, and its application is more versatile. Rather than building large trenches and installing iron walls, initial field trials have shown that nanosized ZVI can be injected directly into the groundwater plume. This minimizes installation costs, which are a major cost component of ZVI PRBs. Little is known about the long-term performance of these nanoparticle/colloidal systems, however, with particular uncertainty surrounding the effect of formation of passivating ferric oxide layers on the outer iron surface.

Pesticide contamination of surface waters, groundwaters, and soils due to their extensive application in agriculture is a growing worldwide concern. Pesticides affect aquatic ecosystems and accumulate in the human body. Approaches to the treatment of pesticide-contaminated soils and waters ranges from conventional methods such as incineration, phytoremediation, and photochemical processes to innovative methods such as ultrasound-promoted remediation and other advanced oxidation processes. Recent studies have shown that many pesticides are susceptible to degradation using ZVI. Preliminary studies in this work on the susceptibility of pesticide degradation using nZVI found that several compounds such as atrazine, molinate, and chlorpyrifos were effectively degraded. Cyclodiene insecticides such as endosulfan, however, were generally very resistant. Molinate appears to be particularly susceptible to degradation by nZVI, and results of more detailed studies on this compound are reported in Chapter 4.

3
Nanoscale ZVI Particles Manufacture and Analytical Techniques

The experimental techniques used in this book are described in this chapter with examination firstly of the method of manufacture of the nanoscale zero-valent iron (ZVI) particles used throughout the study. This is followed by a description of the methods of analysis of both organic compounds (agrochemicals and their degradation products by solid-phase microextraction (SPME) gas chromatography/mass spectrometry (GC/MS) and benzoic acid and p-hydroxybenzoic acid by HPLC) and inorganic compounds (ferrous iron and hydrogen peroxide using colorimetric techniques). A description of the experimental approaches used in detailed studies of molinate and benzoic acid degradation are given, as is a summary of the X-ray diffraction (XRD) approach to determining the nature of inorganic products formed on the surface of the ZVI particles.

3.1. Synthesis of Nanoscale ZVI Particles

Nanoscale ZVI (nZVI) particles were prepared freshly each day by adding 0.16 M $NaBH_4$ (98%, Aldrich) aqueous solution dropwise to a 0.1 M $FeCl_3 \cdot 6H_2O$ (98%, Aldrich) aqueous solution at ambient temperature as described by Wang and Zhang (1997). The synthesis of nZVI was performed under atmospheric conditions. The preparation of solutions involved the following steps: sodium borohydride ($NaBH_4$, 0.6053 g) solids were dissolved in 100 mL of 0.1 M NaOH solution (0.16 M $NaBH_4$ in 0.1 M NaOH solution), and 2.7030 g of $FeCl_3 \cdot 6H_2O$ was dissolved into 100-mL pure water (0.1 M $FeCl_3 \cdot 6H_2O$). $NaBH_4$ solution can be made either in water or NaOH solution, although $NaBH_4$ is unstable in water and can quickly result in a loss of reduction power. Addition of the $NaBH_4$ to the $FeCl_3$ solution in the presence of vigorous magnetic stirring resulted in the rapid formation of fine black precipitates as the ferric iron reduced to Fe^0 and precipitated according to the following reaction:

$$Fe(H_2O)_6^{3+} + 3BH_4^- + 3H_2O \rightarrow Fe^0\downarrow + 3B(OH)_3 + 10.5H_2$$

The particles were washed 3 to 4 times with a 10^{-4} M (pH 4) HCl solution and stored as a 5-mg Fe/mL concentrate at pH 4 and kept in a cooling room ($< 4°C$).

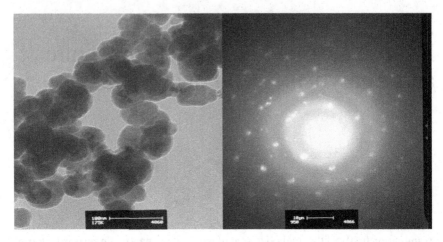

FIGURE 3.1. TEM images of (a) primary ZVI particles (scale bar = 100 nm) and (b) the diffraction pattern (scale bar = 10 μm).

The total amount of ZVI particles produced was 0.2 g assuming that all the soluble Fe(III) is reduced to Fe^0. Although the total amount of iron by estimation of mass balance is 0.2 g, there were likely to be iron losses during the acid-washing step. A series of control tests indicated that the mass of Fe^0 lost due to this process was approximately 5% (Agrawal and Tratnyek, 1996).

3.1.1. ZVI Particle Characterization

Dry particles for particle characterization were obtained by washing the wet precipitates with 10^{-4} M HCl 3 to 4 times, followed by rinsing with pure water, and then separating using a centrifuge at 3000 rpm for 5 min to remove the remaining moisture. The ZVI particles were then quickly frozen using liquid nitrogen and freeze dried under vacuum for more than 20 h. Compared with freeze drying under vacuum, drying under air resulted in the color of Fe particles changing from black to reddish-brown within a few hours, indicating significant surface oxidation. Analysis of the freeze-dried particles by scanning electron microscopy (SEM) and by transmission electron microscopy (TEM) revealed that the primary particle size ranged from 1 to 200 nm with an average size of approximately 50 nm (Figure 3.1a). The presence of a strong diffraction pattern during TEM analysis confirmed that the particles were crystalline (Figure 3.1b). The dried ZVI particles were identified as elemental iron by XRD analysis using a Philip PW 1830 X-ray diffractometer with X'-pert system (Figure 3.2). No other minerals, such as magnetite or maghemite, were identified in the freshly prepared, freeze-dried samples. Single-point Brunauer–Emmett–Teller (BET) analysis by N_2 adsorption (Micromeritics ASAP 2000, GA) determined that the surface area of the particle was 32 m^2/g. The results of the particle sizing and surface area measurements are similar to the results found by other researchers for nanosized ZVI, as shown in Table 3.1.

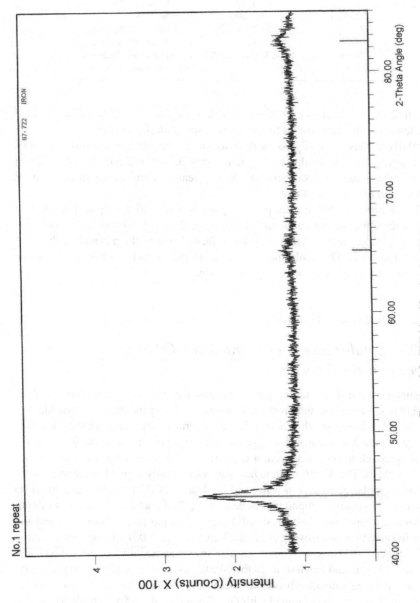

FIGURE 3.2. ZVI identification by XRD analysis.

TABLE 3.1. Properties of various ZVI particles.

Surface area (m²/g)	Size (nm)	References
33.5	1–100	Zhang et al., 1998
31.4	1–200	Choe et al., 2001
27.8–31.8	1–200	This work
0.063	75,000–150,000	Choe et al., 2000 (Commercial grade)
0.038	425,000–850,000	Agrawal and Ttratnyek, 1996 (Commercial grade)

The ZVI particles, however, tend to form larger aggregates, and these may reduce the tendency for the particles to remain in suspension (Figure 3.3).

Malvern MasterSizerE was used to determine the particle size distribution of wet aggregates. The results are shown in Figure 3.4 and indicate the presence of a population of aggregates of around 200-nm mean size and a population of larger aggregates of size >30–40 µm.

In addition to surface area, primary particle size, and aggregate particle size measurements, the zeta potential was measured (using a Brookhaven ZetaPlus particle charge analyzer) to assess the surface charge of the particles at different pHs (Figure 3.5). The results indicate that at low pH the particles have a net positive charge, and at higher pH a net negative charge.

3.2. Analytical Techniques

3.2.1. Solid-Phase Microextraction GC/Mass Spectrometer Detector

Pesticides are difficult to analyze because of the wide range of chemical characteristics, structures, and properties. For example, organochlorine pesticides are classified as nonpolar, while herbicides are generally very polar semivolatiles. This range of polarities creates challenges for effective preconcentration and analysis. The approach to preconcentration used here is that of solid-phase microextraction (SPME). The SPME method has been successfully applied in the analysis of some midpolar pesticides in water and food samples (Natangelo et al., 2002), to a variety of volatile compounds (Bouaid et al., 2001), and in the analysis of both polar and nonpolar analytes from solid, liquid, and gas phases (Boyd-Boland and Pawliszyn, 1996; Matisová et al., 2002; Vereen et al., 2000). The preconcentration of nitrogen-containing pesticides has also been successfully accomplished using SPME (Magdic and Pawliszyn, 1996). SPME is typically used prior to GC analysis and is being increasingly adopted because of its simplicity, low cost, rapidity, and sensitivity when combined with GC (Bouaid et al., 2001). Problems may be encountered, however, when environmental samples contain too many unknown components that compete with the target analytes for absorption by the polymer fiber (Eisert and Levsen, 1995).

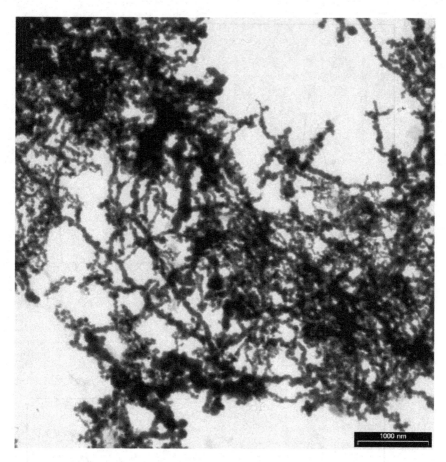

FIGURE 3.3. SEM images of freeze-dried ZVI aggregates manufactured using NaBH$_4$ (scale bar = 1000 nm).

3.2.1.1. SPME Process

The SPME process has two steps: (1) partitioning of the analytes between the sample matrix and a stationary phase that is coated on a fused-silica fiber; and (2) desorption of trapped analytes into the analytical instrument (Beltran et al., 2000; Dugay et al., 1998). Sampling, extraction, and concentration are focused into a single step. In this first step, the coated fiber is exposed to the sample and the target analytes partition from the sample matrix into the coating. Partitioning onto the fiber can be conducted in the sample liquid or headspace. After absorption equilibrium is achieved or after exposure for some defined time, the fiber containing the concentrated analytes is injected into a preheated injector port in a GC system. The fiber is left in the hot injector port for a given period of time (typically 3 min) to provide sufficient time for the analytes to desorb from the fiber. After desorption,

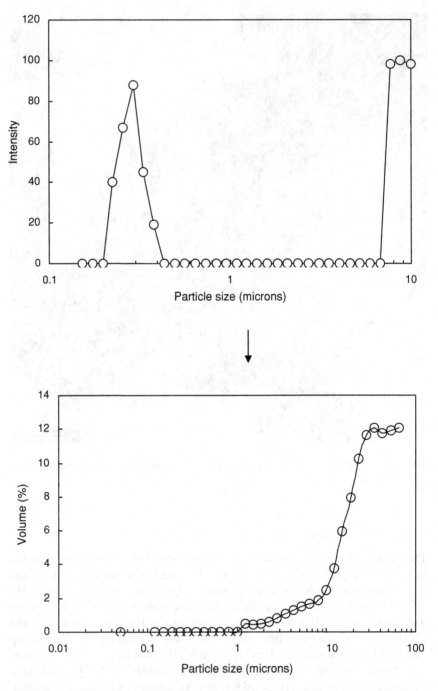

FIGURE 3.4. ZVI particle size distributions determined using Malvern MasterSizerE.

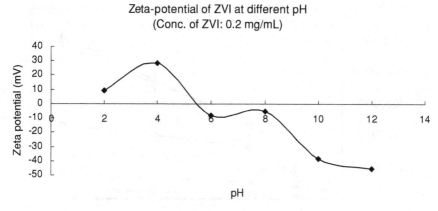

FIGURE 3.5. Zeta-potential of ZVI at different pHs.

the now gaseous compounds are released into the GC column, where they are separated and quantified.

A microextraction fiber coated with 100-μm polydimethylsiloxane (PDMS) was used in all experiments described here because this material has been reported to have a satisfactory extraction efficiency for a variety of compounds including atrazine (Hernandez et al., 2000), several organophosphorus pesticides (Beltran et al., 1998), and organochlorine pesticides (Dugay et al., 1998). High recoveries have been reported (Santos et al., 1996), although the 85-μm polyacrylic acid fiber is better for polar compounds (Buchholz and Pawliszyn, 1993; Magdic et al., 1996). In addition, the PDMS fiber is capable of being used over a high temperature range of 220–320°C, which allows for desorption of higher boiling point semivolatile compounds (Barnabas et al., 1995) and nonpolar compounds (Agilar et al., 1998).

The general SPME procedure used for all experiments in this research was as follows.

1. The coated fibers were conditioned according to the manufacturers' instructions to ensure that any contaminants, which might be present and cause high baseline noise or ghost peaks, were removed prior to use (Boyd-Boland et al., 1996). Preconditioning involved heating the fiber in a GC injector port for 3 h at 260°C. The highest recommended temperature for the PDMS fiber is 260°C (Young and Lopez-Avila, 1996).
2. One milliliter of sample and a small Teflon-coated magnetic stirring bar was placed in a 2-mL glass vial before being sealed with a PTFE-lined septa. The sorption on the Teflon coating of the magnetic stir bar is negligible on the basis of blank and adsorption experiments. While the sample was stirred, the septum was pierced using a stainless steel needle and the PDMS fiber was exposed to the sample for 15 min. Constant rapid stirring was maintained, because the rate at which the extraction process reaches equilibrium is primarily dependent on the rate of mass transfer in the aqueous phase. Although 15 min is insufficient

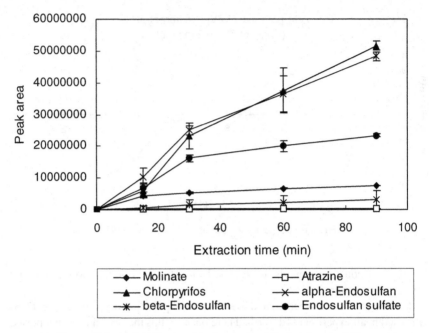

FIGURE 3.6. A plot of extraction time vs. equilibrium for pesticides.

for most pesticides to reach equilibrium concentrations (Figure 3.6), a precise quantification was possible, as this extraction time was kept exactly constant for all experiments conducted. All nZVI batch experiments conducted were reproducible with the relative standard deviations of less than 10%.

The fiber was again withdrawn into the needle and the syringe was removed from the vial. The injector port septum of the GC/MS was pierced with the stainless steel syringe, and the fiber was directly exposed in the hot injector port.
3. Thermal desorption was carried out for 3–5 min at 260°C (considered to be the optimum desorption time for quantification (Ramesh and Ravi, 2001)) and the heating program was initiated. Finally, the fiber was removed from the injector port. Separate vials were used for each experiment as consecutive experiments lead to nonreproducible results.

3.2.1.2. Extraction Technique for Hydrophobic Compounds

The analysis of highly hydrophobic compounds in contaminated soils is challenging because pesticide residues tend to strongly adsorb to soil particles and there is interference from nontarget compounds. The problem has yet to be overcome by many extraction methods. Of the different methods used for extraction of hydrophobic organics from soils (e.g., tumbling, soxhlet, blending, ultrasonic and supercritical fluid extraction) (Capangpangan and Suffet, 1996, 1997; Green, 1996; Lopez-Avila et al., 1994; Noordkamp et al., 1997), ultrasonic extraction was examined in greater detail as it is quick and comparatively easy to do and typically yields satisfactory extraction efficiency (Schneider, 1995). Among solvents such as methanol, acetone, dichloromethane, and hexane, which are commonly used for extraction, n-hexane was chosen as hexane and methanol are the most used solvents for extracting organochlorine pesticide (e.g., endosulfan) residues from environmental samples (Kanatharana et al., 1993).

Using ultrasonic/hexane extraction, centrifugation and direct injection into GC/MS gave a recovery rate of 70% for endosulfan from a contaminated sand mixture. In this study SPME GC/MSD was only used for solution measurement, while GC/MSD was employed for the extraction of water/sand mixture into n-hexane. This is because the 100-μm PDMS nonbonded fiber cannot be used with nonpolar organic solvents. Only fibers that are indicated as bonded can withstand most organic solvents. In addition, glass was used in all the apparatus to reduce poor recovery as organics are readily absorbed to plastic tubes. It was found that without the SPME step (e.g., by direct injection using solvents), the sensitivity was too low for the dilute levels (e.g., 10 ppb) typically found in waterways.

3.2.1.3. GC/MS Analysis

As indicated in Section 3.2.1.1, the fiber was removed from the sample and introduced into the GC/MS injector, where the analytes were thermally desorbed for 3 min and injected onto a (HP-5MS) column in splitless mode with the injector held isothermally at 260°C. For quantification of molinate, the GC/MS injector was operated in the selected ion monitoring (SIM) mode by monitoring the base peak of molinate (m/z 126, 55, 187, 83). The mass spectrometer was tuned to m/z 126, 55, 187, 83 for the electron impact (EI) corresponding to that of perfluorobutyl-amine (PFBTA). In some experiments, by-products were analyzed using a full scan between 50 and 500 m/z.

The temperature program in SIM mode was as follows: the initial temperature was 80°C (1 min), which was increased to 178°C (3 min) at 30°C/min, and then to 250°C (5 min) at 30°C/min, giving a total run time of 14.7 min. In the full-scan mode the initial temperature was 80°C, which was increased to 200°C at 5°C/min, and then to 210°C (3 min) at 5°C/min, followed by to 270°C (3 min) at 20°C/min, giving a total run time of 35 min. For both analytical methods the detector was set at 280°C, helium (pure carrier gas grade) was used as the carrier gas at a flow rate of 1.0 mL/min, and the EI ionization energy was set to 70 eV. All standard curves involved use of five concentrations and were linear, with regression coefficients greater than 0.9995 in all cases. For molinate, the method provided a limit of detection of 10 ng/L. In order to prevent carryover, blanks were run before the next sample extraction. To account for the filter recovery (95%), initial samples (before adding ZVI) were also filtered.

3.2.2. HPLC Analysis of Benzoic Acid and p-Hydroxybenzoic Acid

Benzoic acid (BA) and p-hydroxybenzoic acid (p-HBA) concentrations were quantified by HPLC using a Hewlett-Packard 1100 series HPLC system equipped with a 250 × 4.6 mm Waters Spherisorb ODS-2 5μ column (Alltech, IL). A two-solvent gradient elution consisting of pH 3 water (adjusted with 85% o-phosphoric acid) and acetonitrile (85:15, v/v %) at a flow rate of 1.0 mL/min was used to separate BA and isomers of p-HBA. The p-HBA isomer was quantified at 255 nm and BA was quantified at 270 nm. All standard curves were linear, with regression coefficients of >0.9990 in all cases. The method detection limit for BA and p-HBA were 2.5 μM and 0.1 μM.

3.2.3. Measurement of Ferrous Iron Concentrations

Ferrous ion generation for both the molinate and BA experiments was quantified by monitoring the absorbance of a Fe(II)-bipyridine complex at 522 nm in a manner similar to that described by Voelker (1994). Samples were filtered through a 0.22-μm Millex-GS syringe filter before analysis. The procedure consisted of premixing 0.4 mL of a pH 6 phosphate buffer (0.5 M) and 0.1 mL of bipyridine solution (0.01 M) in a 5-cm spectrometer cell and adding 2 mL of the filtered sample, followed by 0.02 mL of EDTA stock (0.01 M Na_2EDTA). After 60 s, the absorbance was measured at 522 nm.

Ferrous ion generation for the molinate experiments was also quantified by continuously monitoring the absorbance of a ferrozine–Fe(II) complex at 562 nm in a flow injection apparatus similar to that described by Rose and Waite (2002). Analysis of Fe^{2+} evolution from the 1.79 mM ZVI suspension required 1 mM ferrozine, while 4 mM ferrozine was used to measure the total Fe^{2+} released from 5.36 mM ZVI. The higher 4 mM ferrozine was not necessary for the 1.79 mM ZVI and did not affect Fe^{2+} measurement. Samples were continuously pumped through a 0.45-μm glass fiber (GF) filter before injection. For dissolved ferrous measurement, the ZVI samples were filtered using GF filter before combining with ferrozine, while total ferrous iron (adsorbed and dissolved Fe(II)) was measured by mixing with ZVI samples and ferrozine before filtering using GF. Schematics of the approaches to dissolved and total ferrous iron analysis are shown in Figures 3.7 and 3.8.

3.2.4. Measurement of Hydrogen Peroxide (H_2O_2) Concentrations

The concentration of hydrogen peroxide was measured in the absence of organic compounds (e.g. molinate) in these studies. The method used is similar to that developed by Balmer and Sulzberger (1999) and involves the use of the reagent DPD (N,N-diethyl-p-phenylenediamine): (1) 0.4 mL of phosphate buffer solution

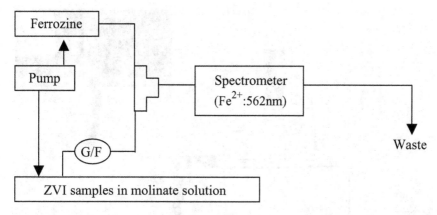

FIGURE 3.7. Schematic of dissolved ferrous iron measurement.

(pH 6.0, 0.5 M phosphate) and 0.1 mL of pure water were premixed in a 1-cm quartz cell; (2) 2 mL of sample solution from which ZVI particles had been removed by filtration was added, followed by the addition of 0.02-mL EDTA solution (0.01 M Na_2EDTA); (3) 0.03 mL of DPD reagent (1% in 0.1 M H_2SO_4) was then added, followed by the addition of 0.03-mL POD (horseradish peroxidase) reagent (about 0.8 mg/mL); and (4) the absorbance was then measured at 551 nm.

3.3. Procedures Used in nZVI-Mediated Degradation Studies

3.3.1. Molinate Degradation

3.3.1.1. Reagents

Molinate (s-ethyl-perhydroazepin-1-thiocarboxylate; 99% purity) was purchased from Alltech Associates (Australia) Pty. Ltd. All chemicals used in this work were analytical reagent grade, and solutions were prepared in ultrapure water (Milli-Q water, Millipore). Molinate solutions were prepared from a 100-ppm stock solution (water solubility of molinate is 880 ppm at room temperature; Hartley and Kidd, 1983).

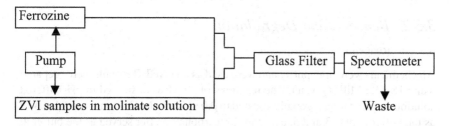

FIGURE 3.8. Schematic of total ferrous iron measurement.

FIGURE 3.9. Typical system of GC/MS with SPME.

3.3.1.2. Experimental Setup

Experiments were carried out at room temperature ($20 \pm 2°C$) under both air and nitrogen atmospheres without pH control (pH $\sim$ 6–7) as well as at fixed pHs of 4 (obtained by HCl addition) and 8.1 (obtained using a 2 mM bicarbonate buffer). Experiments were conducted in 100-mL serum bottles using a total suspension volume of 50 mL under aerobic conditions, while the vials were completely filled (i.e. no headspace) for studies under deoxygenated conditions. ZVI suspensions were prepared from the 5-mg Fe/mL stock suspensions, and molinate was added using a micropipette. Each bottle was either continuously shaken using an orbital shaker at 175 rpm (Hybritech Inc.) or continuously sparged with the indicated gas or mixture, and samples were taken at regular time intervals using a syringe and filtered through a 0.45-μm GF filter before SPME GC/MS analysis. Typical system of GC/MS with SPME is shown in the Figure 3.9.

3.3.2. Benzoic Acid Degradation

3.3.2.1. Reagents

All chemicals were high purity and were used as received. Reagents were prepared using 18-MΩ Milli-Q water. The reactions of nZVI were studied in pH-buffered solutions with an ionic strength adjusted to 0.03 M with NaCl. Benzoic acid served as the buffer at pHs 3 and 5, and 2 mM bicarbonate buffer served as the buffer at pH 8. When necessary, the pH of the solutions was adjusted using 0.1 N HCl or 1 N NaOH. Synthetic humic acid and commercial ZVI powder (>97% purity, particle size 4.5–5.5 μm) were obtained from Aldrich. Benzoic acid, hydroxybenzoic acid, and phenol were obtained from Sigma-Aldrich, and aniline was obtained Ajax FineChem Ltd. Nanoscale ZVI was synthesized as described in Section 3.1.

To compare the nZVI to other forms of $Fe^0_{(s)}$ the following were used: Master Builders granular ZVI (particle size 750–1200 μm; Orica Chemicals, Australia), electrolytic ZVI powder (>98% purity, particle size 100–150 μm; Kanto Chemical Co., Inc, Japan), and Aldrich fine ZVI powder (>97% purity, particle size 4.5–5.5 μm). The specific surface areas of the Master Builders, Aldrich, and Kanto iron and nZVI were 0.71, 0.38, 0.60, and 32 m^2/g, respectively, as determined by BET analysis using a Micromeritics ASAP 2000 instrument.

3.3.2.2. Experimental Setup

All experiments were carried out at room temperature ($20 \pm 2°C$) in 100-mL serum bottles using a total volume of 50 mL. The nZVI particles were kept in suspension by placing the bottles on an orbital shaker table at 175 rpm. To ensure gas exchange, the serum bottles were open to the atmosphere. The solution pH was continuously monitored with a pH electrode throughout each experiment, and adding nZVI usually resulted in an increase during the initial phase of the reaction. To maintain a constant pH value, HCl was added with a 1000-μL Eppendorf pipette as needed during the first hour of the experiments. After the first hour, the pH remained constant. The pH of the suspensions was maintained within ±0.2 units from the initial values during the reaction runs.

To initiate a reaction, nZVI was added from the stock suspensions to buffered solutions containing benzoic acid. Samples were collected at different time intervals in a 5-mL glass syringe and filtered immediately through a 0.45-μL Millipore (Millex AP 20) glass syringe to stop the reaction. Loss of p-HBA on the filter was less than 5%, as determined by comparison of filtered and unfiltered samples collected prior to addition of nZVI. p-HBA was analyzed within 24 h of sample collection.

Experiments also were performed to assess the relative rates of hydroxyl radical oxidation of several probe compounds. These experiments were performed using Fenton's reagent as a source of hydroxyl radical. Experiments were conducted in serum bottles by adding an excess of Fe(II) (i.e., 200 μM) to H_2O_2 (20 μM) at pH 3 with an ionic strength of 0.03 M and 10 mM benzoic acid. Samples collected after 10 min were analyzed for p-HBA by HPLC.

3.4. Experimental Setup Used in ZEA System Studies

Approximately 0.5 g of iron particles (40–70 mesh, 99%; Alfa-Aesar, Ward Hill, MA) were added directly to 10 mL of aqueous solution containing 1.1 mm of 4-chorophenol (4CP) (99%+, Acros) and 0.32 mM EDTA (99%+, Acros, Pittsburgh, PA) and then stirred under room temperature and pressure conditions. Subsequent control runs were conducted in the absence of EDTA and/or under N_2 gas flow. Pentachlorophenol (PCP) (98+, Aldrich, Milwaukee, WI) degradation runs were conducted in a similar fashion. This species is of limited aqueous solubility (at pH 5.5), and the addition of 0.61mM of PCP to 10.0 mL of an aqueous 0.17 mM EDTA solution along with 0.5 g of 40–70 mesh granular ZVI results in a heterogeneous slurry. It was observed that the initial pH of these solutions was 5.5 in all cases. It is known that ZVI corrosion could bring the pH of an aqueous system to >10. However, no attempts were made to regulate the pH of these reaction mixtures, as a result of the system self-buffering at a pH between 5.5 and 6.5. The degradation products consisting of low-molecular-weight organic acids and EDTA is thought to play a large role in the ability of this system to self-buffer.

After the allotted reaction time the reaction mixture products were extracted into 2 mL of ethyl acetate (Optima grade, Fisher Scientific, Pittsburgh, PA) and analyzed on a GC-FID system (Hewlett-Packard 5890A). To ensure that the analytes were in the protonated form the reaction mixture was acidified to pH 2 using sulfuric acid. The GC temperature program used for the analysis of reactants and products had an initial temperature of 60°C for 2 min, followed by a ramp of 10°C/min to a final temperature of 270°C for 5 min. The injection port and flame ionization detector (FID) temperatures were 270°C and 300°C, respectively. All injections were performed in splitless mode using helium as a carrier gas at a rate of 3.35 mL/min. The separation column was an Alltech EC-5 (0.32 mm i.d., 0.25-μm film). Mass spectral analyses were conducted on a Jeol JMS-AX505 HA. Ionization was conducted by EI with a current of 100 μA. The GC separation conditions were identical to those used in GC-FID. Direct probe measurements between 50 and 800 amu were conducted, with an acceleration voltage of 3.0 kV. Reaction kinetic data were performed on a series of separate reaction mixtures.

A stock EDTA (99.9%, J.T. Baker) solution was prepared by dissolving Na_2H_2EDTA in deionized water (Fisher, HPLC grade, New Jersey). The reaction vessel was a 150-mL round bottom flask, with a total suspension volume of 50 mL. BET surface area analysis of the granular ZVI (Aldrich, A.C.S. grade, 20–40 mesh) was performed by Porous Materials, Inc. (Ithaca, NY), and found to be 0.1105 m^2/g. The initial concentration of EDTA was set at 1 mM (unless noted). A pH of 5.5–6.5 was maintained by the reaction mixture without the use of a buffer. Because of the self-buffering nature of the reaction products, no attempts were made to regulate the pH of these reaction mixtures. Experiments were conducted at room temperature ($20 \pm 2°C$) and open to the atmosphere. Reproducible stirring was accomplished using a Bioanalytical Systems controlled-growth mercury electrode apparatus (West Lafayette, IN). Temperature control was maintained by a custom-made, 100-mL water-jacketed vessel with a circulating water bath (Haake GH-D1). The uncertainty in the activation energy measurements was calculated using the propagation of errors method (Bevington and Robinson, 2003).

Concentration of $Fe^{III}EDTA$ was measured on the basis of an HPLC method developed by Nowack et al. (2002). The mobile phase consisted of 92%, 0.02 M formate buffer of pH 3.5 (formic acid sodium salt, 99%, Acros, NJ, and formic acid, 89.5%, Fisher, Fairlawn, NJ), 8% acetonitrile (Fisher, HPLC grade), and 0.001 M tetrabutylammonium bromide (TBA-Br) (Acros, 99+%, New Jersey). The HPLC included a sampling loop of 20 μL with an Alltech (Deerfield, IL) Econosphere C18 column, (length 150 mm, diameter 4.6 mm, 5-μm packing diameter), with a corresponding guard column. Detection was by means of a UV absorbance detector (Hewlett-Packard 1050 series) at a wavelength of 258 nm. Control experiments were run in the absence of ZVI where the $Fe^{III}EDTA$ complex was created by adding equal millimolar amounts of $Fe(NO_3)_3$ (Fisher, 99+%, NJ) to the EDTA sample prior to HPLC analysis. Controls showed that neither EDTA nor Fe^{3+} salts alone gave UV absorbance at 258 nm.

Nonvolatile aqueous phase reaction products were analyzed using a Micromass Quattro II mass spectrometer equipped with an electrospray ionization probe, two quadrapole analyzers, and a hexapole collision cell. All samples were passed through a 0.4-μm filter and delivered into the source at a flow rate of 5 μL/min using a syringe pump, analysis was done in both negative and positive ion mode, and MS/MS was used for positive identification of peaks. A potential of 2.5 kV was applied to the electrospray needle. The sample cone was kept at an average of 15 V and the counterelectrode, skimmer, and radio frequency lens potentials were tuned to maximize the ion beam. Argon was used as a collision gas during daughter analysis.

CO_2 headspace analysis was performed on the reaction mixture using a Hewlett-Packard series II 5890 gas chromatograph equipped with a PoraPlot Q capillary column (25m $\times$ 0.32 mm) (Chrompack, Middelburg, The Netherlands). The injector temperature was 150°C, and that of the GC/MS interface was 230°C. Separations were achieved using a temperature program of 60°C (4 min) with ramping to 220°C at a rate of 25°C/min. Samples (1.0 μL) were introduced using an autoinjector.

The presence of oxidizing species was qualitatively detected using the thiobarbituric acid-reactive substances (TBARS) assay (Bucknall, 1978; Gutterigded, 1987). The TBARS assay is a nonselective assay that is able to detect many oxidizing species present, including hydroxyl radicals and other high-valent iron-oxo species. Damage to the deoxyribose by the oxidizing species through a hydrogen abstraction mechanism results in the release of thiobarbituric acid-reactive material that can be detected at 532 nm. Deoxyribose (3.18 mM, 99%, Acros) was incubated in a solution of 2.39 mM EDTA and 0.100 g ZVI for 30 min, after which a TBARS chromogen was produced by allowing the oxidized deoxyribose products (10 mL) to react with 5 mL of 104 mM thiobarbituric acid (98%, Acros), heated in a 50°C water bath for 15 min. The TBARS product was detected by optical absorbance at 532 nm on a Hewlett-Packard 8432 diode array spectrophotometer. Controls consisted of a run in which EDTA and deoxyribose were absent.

3.5. Determination of ZVI Surface Products by XRD

In addition to confirmation of the identity of the primary reactant (ZVI), XRD analysis was also used to examine the nature of mineral products formed on the ZVI surface during degradation studies. Details of the procedures used are given below.

3.5.1. Measurements in the Presence of Molinate

XRD analysis was performed using a Philip PW 1830 X-ray diffractometer with X′-pert system. Samples of ZVI particles (from studies containing 1.79 mM ZVI) were collected after reaction with molinate (100 ppb) over 3 h in both unbuffered and buffered systems (i.e., 2 mM NaHCO₃, 20 mM NaCl, pH 8.1). All the ZVI particles were separated after a reaction time of 3.25 h, and then moisture was removed by centrifugation (3000 rpm for 10 min, at 15°C), freezing at −84°C using liquid nitrogen, followed by freeze drying for 20 h.

3.5.2. Measurements in the Absence of Molinate

Four samples in the absence of molinate were obtained for XRD analysis as follows. Samples 1 and 3 were collected from 10.7 mM ZVI suspensions in both Milli-Q water and 2 mM $NaHCO_3$/20 mM NaCl solutions (pH 8.1) after 3 h of reaction, while samples 2 and 4 were collected after 3 h of reaction from 10.7 mM suspensions in both Milli-Q water and 2 mM $NaHCO_3$/20 mM NaCl solutions (pH 8.1) to which H_2O_2 (0.33 mM) had been added. After 3 h, all the ZVI particles were collected and rinsed with Milli-Q water 2 times. Dried ZVI was obtained as described earlier.

4
Oxidative Degradation of the Thiocarbamate Herbicide, Molinate, Using Nanoscale ZVI

4.1. Introduction

The herbicide molinate [S-ethyl hexahydro-1H-azepine-1-carbothioate] is used extensively to control germinating broad-leaved and grass weeds during rice production. It is a moderately persistent ($\tau_{1/2} = 21$ days) (Weber, 1994) and highly soluble (solubility = 880 ppm (Hartley and Kidd, 1983)) herbicide that is often detected as a contaminant in rainwater (Charizopoulos and Papadopoulou-Mourkidou, 1999; Sakai, 2003; Suzuki et al., 2003), lakes (Nohara et al., 1997; Sudo et al., 2002), rivers (Cerejeira et al., 2003; Coupe et al., 1998; Crepeau and Kuivila, 2000; Paune et al., 1998), and estuaries (Oros et al., 2003) worldwide. Molinate is one of the most frequently detected pesticides/herbicides exceeding drinking water guidelines and guidelines for the protection of aquatic ecosystems in southern Australian irrigation networks and rivers (Australian State of the Environment Committee, 2001). Molinate has a high cancer hazard factor (Gunier et al., 2001; Kelly and Reed, 1996), primarily as a consequence of possible exposure to contaminated air and dust (Lee et al., 2002); the long-term effects of its presence in aquatic ecosystems are of concern (Coupe et al., 1998; Crepeau and Kuivila, 2000).

As noted in Chapter 2, molinate was found to be especially susceptible to degradation by zero-valent iron (ZVI) in screening studies conducted on a range of agrochemicals (Joo et al., 2002b). Therefore, molinate was selected for detailed investigation in order to examine the degradation mechanism and improve our understanding of how the removal rate could be enhanced, particularly in field situations.

4.2. Results

4.2.1. Effect of the Presence of Air/Oxygen

The degradation of 100 ppb molinate was examined both under deoxygenated conditions and in the presence of air and 100% oxygen. As shown in the Figure 4.1, molinate showed little removal in pH 4 solutions in the absence of oxygen, while

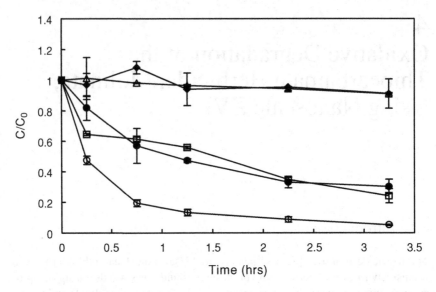

FIGURE 4.1. Comparison of 100 ppb molinate removal at pH_0 4, 10.7 mM ZVI when sparging with N_2 (♦), zero-grade air (●), or 100% O_2 (○) or shaking in the presence of air (□). Also shown is a control in which a ZVI-free solution containing 100 ppb molinate is sparged with N_2 (△) (Joo et al., 2004).

70% removal was observed over 3 h when the sample was sparged with air. Removal was further enhanced when the suspension was sparged with pure oxygen. Vigorous shaking of the vials under atmospheric conditions resulted in a similar degree of removal as sparging with zero-grade air. A control with N_2 sparging showed no removal by volatilization over 3 h (Figure 4.1).

4.2.2. Effect of Molinate and ZVI Concentration

For any particular ZVI concentration, molinate degradation under oxic conditions exhibits pseudo-first-order decay with both a linear dependence of $\ln(C/C_0)$ on time and an independence of the observed pseudo-first-order rate constant (k) on initial molinate concentration (Figures 4.2 and 4.3). As can be seen from Figure 4.3, a linear dependence of rate of molinate degradation on ZVI concentration (measured as total Fe^0) is observed, suggesting an overall rate expression of the form

$$\frac{d\,[\text{molinate}]}{dt} = -k\,[\text{molinate}]\,[\text{ZVI}],\qquad(4.1)$$

where the second-order rate constant $k = 1.8 \times 10^{-2}\ \text{M}^{-1}\text{s}^{-1}$.

The effect of different ZVI concentrations on the degradation of molinate at two different concentrations (5 and 100 ppb) is shown in Figure 4.3. As can be seen in this figure, the molinate disappearance rate is strongly dependent on ZVI

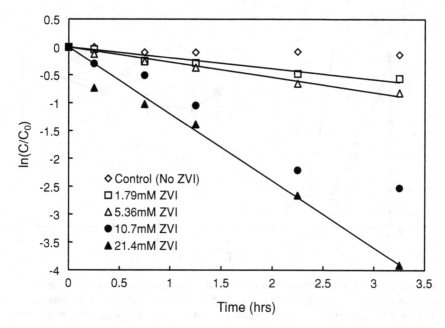

FIGURE 4.2. Molinate degradation in unbuffered solutions ($pH_{initial} \approx 6.4$) for different ZVI concentrations and $[Molinate]_0 = 100$ ppb (Joo et al., 2004).

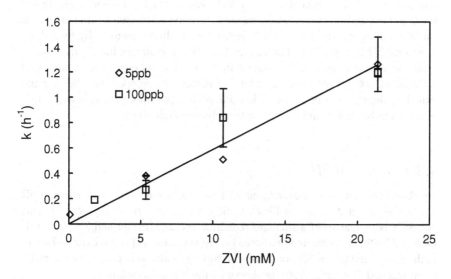

FIGURE 4.3. Dependence of pseudo-first-order rate constant (k) for molinate degradation on ZVI concentration (as quantified by total Fe^0 concentration) for initial molinate concentrations of 5 and 100 ppb in unbuffered solutions ($pH_{initial} \approx 6.4$) (Joo et al., 2004).

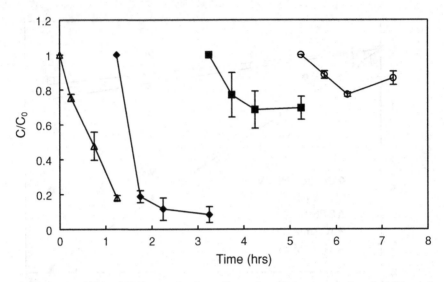

FIGURE 4.4. Ability of ZVI to maintain molinate degradation on continued addition of aliquots of molinate. Conditions: pH = 4, $[ZVI]_0 = 10.7$ Mm; addition of molinate at 0, 1, 3 and 5 h resulted in increase in molinate concentration in reaction medium of 100 ppb (Joo et al., 2004).

concentration. As the concentration of ZVI increases, the degradation rate increases for both high and low initial molinate concentrations. The results of control runs (no ZVI) also, shown in Figure 4.2, reveal that molinate removal by hydrolysis at neutral pH is negligible. The capacity of ZVI to maintain the degradation of molinate was examined and is shown in Figure 4.4. As can be seen, a second addition of 100 ppb of molinate after 1 h is rapidly degraded, but the rate and extent of degradation of further additions of 100 ppb molinate at 3 and 5 h is observed to be significantly less than that observed initially.

4.2.3. Effect of pH

As observed in other studies, adding ZVI to water resulted in an increase in pH and a concurrent decrease in Eh over the experimental runs (Figure 4.5). This behavior is indicative of a standard reduction process (Elliott and Zhang, 2001; Orth and Gillham, 1996). In unbuffered Fe/H_2O systems, pH rise is expected under both aerobic and anaerobic conditions because of aqueous corrosion of the metal (Agrawal and Tratnyek, 1996), as shown by the following reactions:

$$2Fe^0 + O_2 + 2H_2O \rightleftarrows 2Fe^{2+} + 4OH^- \text{ (aerobic condition)}$$
$$Fe^0 + 2H_2O \rightleftarrows Fe^{2+} + H_2 + 2OH^- \text{ (anaerobic condition)}$$

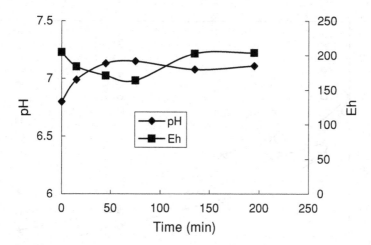

FIGURE 4.5. Variation in pH and Eh on ZVI-mediated degradation of molinate in unbuffered solutions (pH_0 6.8, 1.8 mM ZVI, 100 ppb molinate).

In reaction runs with 1.8 mM ZVI, the pH rose from an initial value of 4 to 7.4 after 3 h of reaction. In solutions buffered to around pH 8 with 2 mM $NaHCO_3$, little change in pH was observed over a 3-h reaction time.

As can be seen from the comparative results in Figure 4.6, molinate degradation occurs at a slower rate in a pH_0 8.1 (adjusted with 2 mM $NaHCO_3$) carbonate-buffered solution than observed in an initial pH_0 4 solution. Significant degradation is still observed at pH 8.1, however, with approximately 60% of 100 ppb molinate degraded after 150-min reaction with 21.4 mM ZVI (Figure 4.6) (c.f. approximately 65% removal at pH 4 for similar reactant concentrations). The rate of molinate removal appears to slow down markedly at times greater than 150 min, while continued degradation is apparent at pH 4.

Increasing the concentration of the ZVI at pH 8.1 does not appear to significantly improve the degradation rate (Figure 4.7). At these higher pH values, oxide and hydroxide coatings develop, which hinders access to the Fe^0 surface (Dombek et al., 2001). The influence of pH on the degradation rate is probably the result of both the available iron surface area for reaction and the competition by other solution components (such as carbonate) for reactive species. According to MacKenzie et al. (1995), a 10-fold increase in aqueous alkalinity reduced the reaction rate by threefold.

4.2.4. Ferrous Iron Generation

4.2.4.1. Ferrous Iron Generation in pH_0 4 Solution

The initial product of ZVI oxidation is ferrous iron. Results of ferrous iron concentration measurement immediately after adding the ZVI particles to pH 4 solutions are shown in Figure 4.8. Immediately after addition of ZVI to water, concentrations

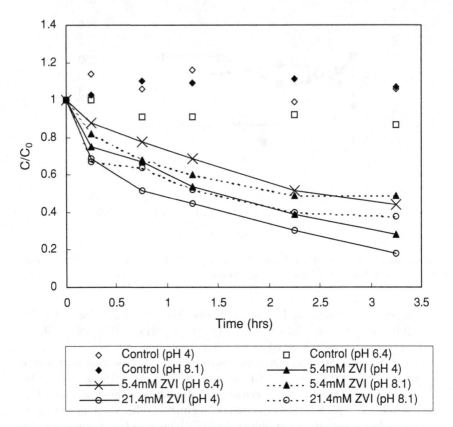

FIGURE 4.6. Comparison of molinate degradation at different starting pHs for [molinate]$_0$ = 100 ppb and [ZVI]$_0$ concentrations of 5.4 and 21.4 mM (Joo et al., 2004).

of Fe(II) ranging from 0.07 to 0.14 mM are detected depending on the amount of ZVI added, and these concentrations subsequently decrease over time. The initial concentrations of Fe(II) observed in solution increase with increasing ZVI concentration, with 72 μM Fe(II) produced on addition of 0.89 mM ZVI, 120 μM produced on addition of 1.79 mM ZVI, and 140 μM produced on addition of 2.68 mM ZVI. An Fe(II) concentration of 140 μM appears to be the maximum achievable as higher concentrations of ZVI do not lead to a greater concentration of Fe(II) in solution.

The subsequent loss of Fe(II) from solution occurs over the ensuing 60–90 min and may result from adsorptive or oxidative processes though the rate of oxygen-mediated oxidation of ferrous iron at pH 4 ($\tau_{1/2} \approx 2$ years) is substantially slower than the rate of removal observed (Stumm and Morgan, 1996).

While the concentration of ferrous iron in solution at pH 8.1 has not been measured, much lower concentrations than observed at pH 4 are to be expected as a result of both the relatively rapid rate of Fe(II) oxidation at pH 8.1 ($\tau_{1/2} \approx 3-4$ min)

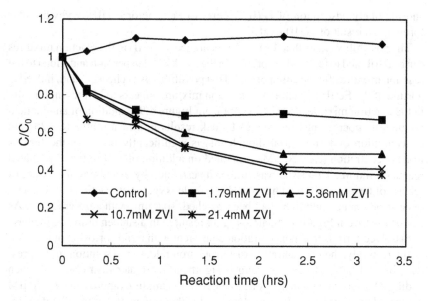

FIGURE 4.7. Molinate degradation in a 2 mM bicarbonate-buffered solution at various ZVI concentrations (100 ppb molinate; pH_0 8.1).

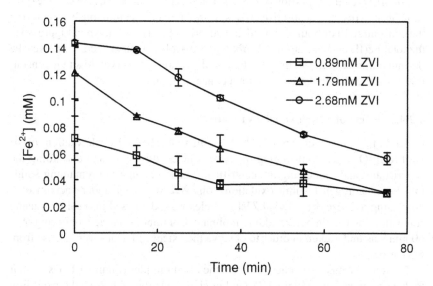

FIGURE 4.8. Dissolved ferrous iron concentrations in pH_0 4 solutions containing 0.89, 1.79, and 2.68 mM ZVI (Joo et al., 2004).

and the likely adsorption of Fe(II) species to Fe^0 and/or Fe(III) oxyhydroxides formed as a result of Fe(II) oxidation.

The possibility exists that the rapid appearance of Fe(II) in the reaction mixtures can be attributed to ferrous iron present in the stock ZVI suspension and added to the reaction mixture after addition of Fe^0. This possibility was checked, and, indeed, a portion of the Fe(II) present in the reaction mixture appears to be present initially in the reaction mixture. As can be seen from Figure 4.9a, addition of an identical volume of supernatant from the ZVI stock results in a measurable (ca. 25 μM) concentration of ferrous iron though this is significantly less than the ferrous iron concentration (ca. 120 μM) generated on addition of ZVI to the oxygenated reaction mixture. This issue was further investigated by examining the extent of release of ferrous iron on addition of ZVI to a deoxygenated reaction mixture and, once steady-state conditions had been reached, sparging with zero-grade air. As can be seen from Figure 4.9b, air sparging resulted in a sudden dramatic increase in dissolved ferrous iron concentration consistent with rapid corrosion of the ZVI.

Interestingly, the concentration of ferrous iron released to solution in the presence of ZVI but under N_2 sparging is significantly greater than observed when adding ZVI supernatant in air alone to the reaction mixture (approximately 70 μM compared to 30 μM from Figures 4.9a,b). It is thus apparent that the ZVI does release some ferrous iron to the reaction mixture that was present in adsorbed form in the ZVI stock. However, at least a doubling in ferrous iron concentration is observed on introduction of oxygen, presumably as a result of rapid corrosion of the ZVI.

Confirmation that a portion of ferrous iron was retained on the ZVI surface (presumably as a result of adsorptive effects) was obtained by continuously monitoring the concentration of both dissolved and adsorbed Fe(II). As shown in Figure 4.10, the total Fe(II) concentration (adsorbed plus dissolved) remains constant over the 60-min duration of the study. The dissolved fraction is observed to attain concentrations somewhat lower than the total concentration.

4.2.4.2. Effect of Use of Dried ZVI Particles

Dried ZVI particles were obtained by washing twice the ZVI stock solution (pH 4) with Milli-Q water, centrifuging the wet particles, decanting and discarding the supernatant, and immersing the centrifuge tube containing the remaining solids into liquid nitrogen to limit oxidation, followed by drying under vacuum over 20 h using a freeze drier. Dried ZVI particles were dispersed into solution using sonication for 10 min before adding molinate. Continuous online monitoring over 60 min was undertaken in order to observe the extent of ferrous iron release from the particles.

The total ferrous iron generated from the dried samples (Figure 4.11) is similar to that observed from colloidal ZVI although the fraction of dissolved ferrous iron from dried ZVI is significantly less than observed for the particles that had been maintained wet. Possibly consistent with the lower extent of ferrous iron release from dried ZVI particles is a smaller increase in pH on addition of the particles to solution compared to that observed for the wet particles.

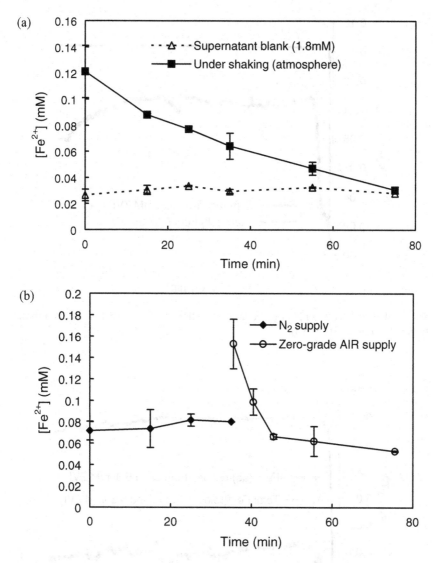

FIGURE 4.9. (a) [Fe^{2+}] in the reaction mixture on addition of a ZVI solution and an equivalent volume of the ZVI supernatant. Conditions: pH$_0$ = 4, [ZVI]$_0$ = 1.8 mM (Joo et al., 2004). (b) [Fe^{2+}] in the reaction mixture after ZVI addition to a nitrogen-sparged suspension with subsequent air sparging. Conditions: pH$_0$ = 4, [ZVI]$_0$ = 1.8 mM (Joo et al., 2004).

The results in Figure 4.12 indicate that there is a somewhat lower rate of molinate degradation for both 1.8 and 5.4 mM ZVI using the dried particles compared with that using the colloidal particles, presumably because of the increased tendency for formation of an outer oxidized layer that inhibits the rate of molinate degradation. While the dried particle results suggest that the overall oxidation process is retarded

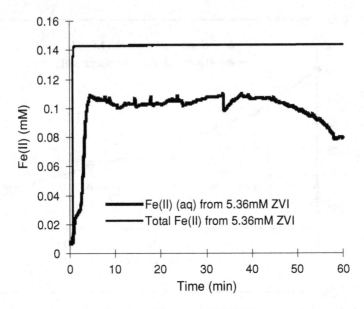

FIGURE 4.10. Fe(II) generation on addition of colloidal ZVI to reaction medium in un-buffered solution ($pH_0 \sim 6.4$).

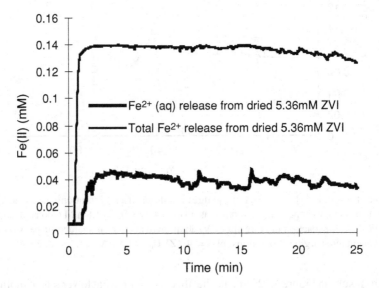

FIGURE 4.11. Fe(II) generation on addition of dried ZVI to reaction medium in unbuffered solutions ($pH_0 \sim 6.4$).

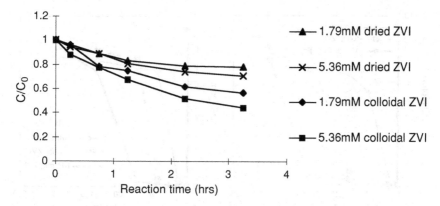

FIGURE 4.12. Comparison of molinate degradation using ZVI concentrate and dried ZVI (100 ppb molinate; $pH_0 \sim 6.4$) (Feitz et al., 2005b).

by surface oxidation, residual oxidizing capacity remains in the particles despite oxidation of the outer surface. The formation of a passivating layer of iron oxide on oxidation of Fe^0 has been described by Davenport et al. (2000), who showed that a layer with properties similar to those of γ-Fe_2O_3 and Fe_3O_4 formed on the particle surface on exposure to oxygen.

Ferrous iron released from ZVI in the presence of 2 mM bicarbonate (pH 8.1) was monitored using the online continuous system. It was observed that the concentrations of Fe^{2+} released from ZVI were much less than for the unbuffered suspensions for both dried and colloidal ZVI, with concentrations for dried and colloidal 5.4 mM ZVI of around 20 μM and 35 μM, respectively (Figure 4.11 and 4.13). The dried ZVI revealed much less Fe(II) release at pH 8.1 than at pH 6.4

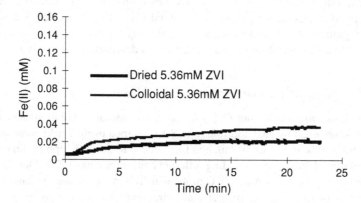

FIGURE 4.13. Dissolved Fe(II) generation from colloidal and dried ZVI at pH 8 in bicarbonate-buffered system.

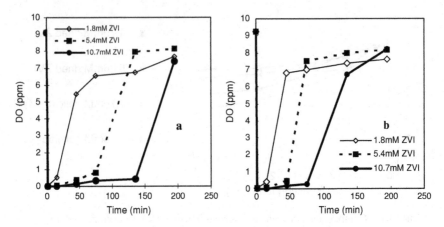

FIGURE 4.14. Dissolved oxygen variations during molinate degradation at (a) pH_0 4 and (b) pH_0 8.1.

(40 μM Fe(II) at pH 6.4. vs. 20 μM Fe(II) at pH 8.1) and presumably reflects both an increased tendency for formation of an iron oxide surface coating and an increased affinity of Fe(II) for this coating at higher pH.

As indicated earlier, there is a general slowing down in the reaction rate at higher pH in the bicarbonate-buffered system although little difference in the initial removal rates for pH_0s ranging from 4 to 8.1 was observed (Figure 4.6). Note that for every mole of oxygen initially consumed (0.26 mM at saturation)—assuming that Fe^0 is in excess—between 2 and 4 mol of H^+ ions are consumed leading to a large rise in pH. The rise in pH is tempered by the diffusion of atmospheric CO_2 into the suspension, and the ensuing carbonate buffering ensures that the maximum pH does not exceed 8.1. Within 15 min, the pH reaches approximately 7 regardless of whether the initial pH is 4 or 6.4 and reaches 7.4 for both experiments after 3 h. There is no change in the pH when the initial pH of the suspension is 8.1. The rapid alignment of pHs, independent of the starting pH over the range studied, is the most likely explanation for the similarities in degradation rates (Figure 4.6).

4.2.5. Effect of Dissolved Oxygen (DO)

The DO concentrations during ZVI corrosion were measured. Solutions with intial pHs of 4 (obtained by HCl addition) and 8.1 (obtained using 2 mM bicarbonate) were saturated with air for 10 min and then spiked to give 100 ppb molinate in a total volume of 50 mL. After adding different concentrations of ZVI, DO changes were monitored using an Au/Ag DO probe with Teflon membrane. As can be seen in Figure 4.14, the dissolved oxygen concentration decreased to zero immediately after addition of ZVI and then increased to >70% saturation within 40–200 min depending on the ZVI concentration and initial pH. The increase in DO is a result of oxygen influx from the atmosphere. Long lag times are observed for higher ZVI

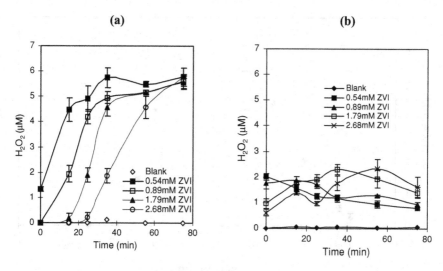

FIGURE 4.15. H_2O_2 generation on addition of various amounts of ZVI at pH_0 4 solutions (a) and pH_0 8 solutions (b).

doses because of a greater capacity for the ZVI surface to consume O_2 (Figure 4.14a,b). These effects are more distinctive at low pH.

4.2.6. Hydrogen Peroxide Generation

Hydrogen peroxide was monitored since it is considered to be a possible product of the reduction of oxygen. The method used is similar to that developed by Balmer and Sulzberger (1999) and involves the use of the reagent DPD (N,N-diethyl-p-phenylenediamine). In the data presented here, no attempt was made to minimize the interaction of Fe(II) with peroxide, as was done by Bader et al. (1988). Later attempts to assess the potential for artifacts associated with the presence of Fe(II) indicated some interference from Fe(II), which could be controlled by adding bipyridine and EDTA. However, the concentrations and trends were not significantly different from those observed in the absence of the complexing agents. Samples were filtered through 0.22-μm filters prior to analysis. At different initial pHs (e.g., pH_0 4 and pH_0 8.1) and constant time intervals, H_2O_2 generation was measured as a function of ZVI concentrations (Figure 4.15). Figure 4.15a shows the hydrogen peroxide generation on addition of ZVI to aerobic pH 4 solutions, with 5–6 μM H_2O_2 produced over extended time intervals of about 1 h or more over a range of ZVI concentrations. Although H_2O_2 was not measured under anaerobic condition, almost no H_2O_2 is expected as indicated by Zečević et al. (1989), who observed that on the oxide-free surface, very little H_2O_2 is formed (less than 0.5% of the total reduction current). Interestingly, steady state concentrations of this magnitude were reached at lower Fe^0 doses more quickly than at higher doses of iron, which would be due to more efficient O_2 scavenging by Fe^0. As the

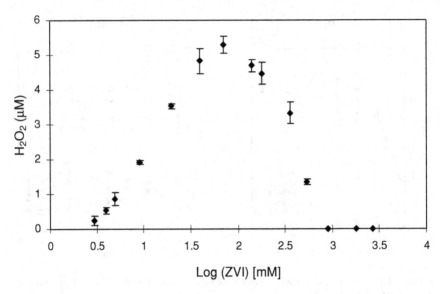

FIGURE 4.16. Instant (initial) generation of H_2O_2 as a function of ZVI at pH 4.

dosage of ZVI was increased, a distinct lag was observed before H_2O_2 could be detected, but with concentrations subsequently increasing to the 5–6 μM level. The detection of micromolar concentrations of H_2O_2 suggests that superoxide is the likely intermediate or the formation of H_2O_2 could be from the protonation of O_2^{2-}, which is the reduction of O_2, i.e.

$$2ZVI = 2Fe^{2+} + 4e^- \qquad (4.2)$$

$$2O_2 + 4e^- = 2O_2^{2-} \qquad (4.3)$$

$$2O_2^{2-} + 4H^+ = 2H_2O_2 \qquad (4.4)$$

The initial absence of H_2O_2 at higher ZVI doses at pH_0 4 is consistent with the rapid reaction of H_2O_2 with Fe^{2+} released during the corrosion process, i.e. the Fenton reaction:

$$Fe^{II} + H_2O_2 = Fe^{III} + OH^\bullet + OH^- \qquad (4.5)$$

k (at pH 4) = 586 $M^{-1}s^{-1}$ (Wells and Salam, 1968),

which results in the production of ferric iron and highly oxidative hydroxyl radicals.

The initial (at time 10 s) peroxide measured (at pH 4) as a function of ZVI revealed that peroxide concentration increased with ZVI dose at relatively low concentrations (a maximum of 5.3 μM H_2O_2 was generated from 70 μM ZVI) followed by a decrease at comparatively higher ZVI doses (Figure 4.16). The maximum initial behavior could be due to the surface total oxygen concentration and the initial Fe(II) concentration in the suspension.

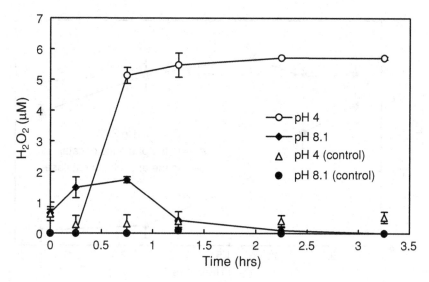

FIGURE 4.17. Comparison of H_2O_2 concentrations in pH 4 (o) and pH 8.1 (♦) solutions containing 1.79 mM ZVI. Possible H_2O_2 concentrations resulting from addition of stock solution are given for pH 4 (▲) and pH 8.1 (●) (Joo et al., 2004).

Small but significant concentrations of hydrogen peroxide (1–2 μM) were detected at pH 8.1. The initial concentrations were slightly higher than those at pH 4 (Figure 4.15b) and again inversely proportional to the ZVI concentration. Unlike at pH_0 4 where H_2O_2 was not detected initially except at very low ZVI concentrations, small amounts of H_2O_2 were measured at 10 s for ZVI concentrations ranging from 0.54 to 2.7 mM at pH_0 8.1 (Figure 4.15b). The difference is probably due to the rapid oxidation of ferrous to ferric by O_2 at higher pH (Scherer et al., 1998), outcompeting reaction via the Fenton reaction (Equation 4.5).

Figure 4.17 shows the results of kinetic experiments conducted at pHs of 4 and 8.1 over longer reaction times. As can be seen, the hydrogen peroxide concentration in pH 4 solutions was quite stable for extended periods of time once a plateau had been reached. In comparison, at pH 8.1, while hydrogen peroxide was certainly still detectable, its concentration peaked (at about 2 μM) within about 30 min of adding the ZVI to the solution and then decreased to immeasurable levels in the ensuing 2–3 h.

4.2.7. Catalase and Butanol Competition

That hydrogen peroxide plays a key role in the degradation of molinate is suggested by examination of the effect of addition of catalase, an enzyme that induces the rapid degradation of hydrogen peroxide, to the experimental system. As can be seen in Figure 4.18, no degradation was observed in the presence of catalase. While

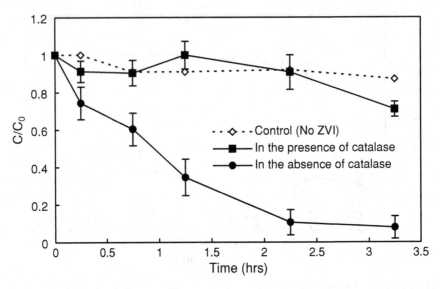

FIGURE 4.18. Effect of ZVI-mediated degradation of molinate of 60 ppm catalase. Conditions: [Molinate]$_0$ = 100 ppb and [ZVI]$_0$ = 10.7 mM (Joo et al., 2004).

this result is suggestive, it is not definitive of a key role for hydrogen peroxide. Other oxidants such as the powerful hydroxyl radical could be involved that may well be scavenged by compounds such as catalase. Indeed, as can be seen from the results shown in Figure 4.19, little degradation of molinate is observed in the presence of 1-butanol, an effective hydroxyl radical scavenger.

4.2.8. Degradation By-products

Reaction intermediates are usually of low stability and undergo fast degradation, making detailed study of intermediate species a difficult task. Moreover, they occur at low quantities, sometimes only in trace amounts, and are difficult to isolate and identify. As indicated by Pichat (1997), most intermediates can be identified only when the initial substrate concentration is raised by about two orders of magnitude. High concentrations (10 mg/L) of molinate were degraded using ZVI in order to isolate and identify the intermediate products. By-products were identified using GC/MS analysis. Since the by-product peaks were relatively small compared with that of molinate, all the ZVI particles in suspension containing molinate (10 ppm) after the final reaction time (3.25 h) were removed in order to clearly identify the intermediate peaks. ZVI particles were separated from the suspension using magnetic separation. Two milliliters of hexane was then added to the wet ZVI particles, which were sonicated for 20 min and, using centrifugation, at 2000 rpm for 5 min. The extracted supernatant was analyzed using GC/MS-EI (TIC). An

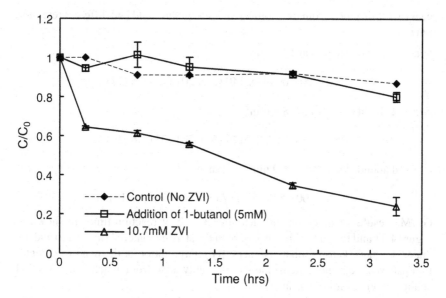

FIGURE 4.19. Effect on ZVI-mediated degradation of molinate; 5 mM 1-butanol. Conditions: $[Molinate]_0 = 100$ ppb and $[ZVI]_0 = 10.7$ mM; butanol study undertaken at pH_0 4 (Joo et al., 2004).

example chromatogram for ZVI/O_2 showing molinate and keto-molinate peaks as a function of retention time is given in Figure 4.20.

Qualitative analysis using SCAN mode (total run time = 35 min) was performed and mass spectra were collected in the total ion current (TIC) mode to characterize each fragment. Two peaks, which exhibited retention times of 17.6 and 18.8 min, were closely investigated since these peak areas increased as molinate concentration decreased during the reaction. The mass fragments of these

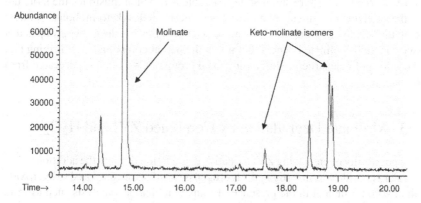

FIGURE 4.20. GC/MS-EI (TIC) obtained from ZVI extracted after the reaction time of 3.25 h.

peaks were different from the molinate mass fragments and had different retention times:

m/z and abundance (%) at 17.6 min:

$$112\ (100),\ 69\ (44),\ 140\ (25),\ 201\ (45),\ 56\ (17)$$

m/z and abundance (%) at 18.8 min:

$$112\ (100),\ 69\ (22),\ 140\ (65),\ 201\ (35),\ 56\ (17)$$

m/z and abundance (%) at 14.9 min (molinate):

$$126\ (100),\ 55\ (39),\ 187\ (37),\ 83\ (16),\ 98\ (14)$$

GC/MS results obtained on samples after 3.25 h of reaction time are shown in Figure 4.21 and reveal the presence of both the starting material (molinate) and the major products, keto-molinate isomers. By interpretation of the mass fragmentation patterns, keto-molinate isomers presumably arise from attack at the N-alkyl chain (azepine ring) of molinate.

The mass spectra of these peaks do not have the hexahydroazipine isocyanate ion (m/z: 126), the base peak of molinate, and yield oxygenated hexahydroazipine isocyanate ions (m/z: 140) and weak molecular ions (m/z: 201). The major products, keto-molinate isomers, are formed by oxygen addition to partially oxidized molinate. This observation was consistent with the findings of Konstantinou et al. (2001a), who identified keto-molinate isomers from the photocatalytic degradation of molinate over aqueous TiO_2 suspensions.

The process is rationalized as hydrogen abstraction by surface-bound hydroxyl radicals occurring preferentially on the N-alkyl chain, resulting in rapid decomposition, particularly in the case of aliphatic thiocarbamates (Konstantinou et al., 2001b).

The likely fragmentation processes leading to the molinate and keto-molinate mass/charge patterns are summarized in Table 4.1, which illustrates the structure of the main mass fragments of molinate and intermediates (keto-molinate isomers) with the functional group lost. As can be seen in Table 4.1, the C=O group is lost twice in keto-molinate isomers (but not in the parent compound), confirming that the keto-molinate structure has a functional group of C=O that is generated from oxidation.

4.3. Molinate Degradation by Combined ZVI and H_2O_2

The possibility of enhancing ZVI oxidative degradation through the addition of hydrogen peroxide was investigated. Hydrogen peroxide is decomposed to hydroxide and hydroxyl radical in the presence of transition-element catalysts such as iron. Hydroxyl radicals are strong, nonselective oxidants that react with most organic

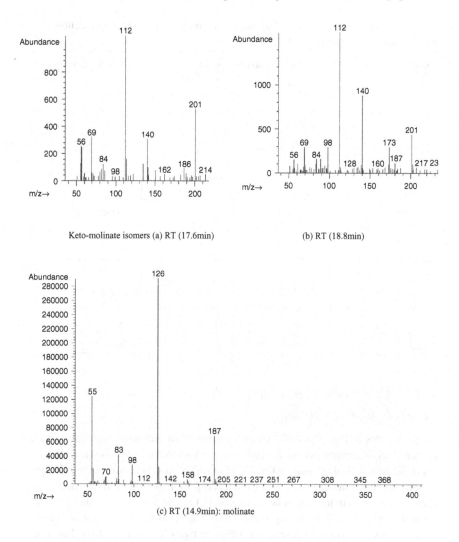

FIGURE 4.21. Mass fragments for each peak on keto-molinate (a,b) and molinate (c). (a) Reaction time (RT) 17.6 min; (b) RT 18.8 min; (c) RT 14.9 min.

compounds at rates near the diffusion-controlled limit of 10^{10} $M^{-1}s^{-1}$. Dowling and Lemley (1995) found that organophosphate insecticides and their breakdown products are susceptible to degradation via hydroxyl radical attack, and preliminary studies showed that the addition of peroxide to nanoscale ZVI (using a 1.6:1 ratio of Fe(0) and H_2O_2) significantly increased the removal rate of organophosphorus insecticides such as chlorpyrifos (>90% degraded within 15 min).

TABLE 4.1. Interpretation of mass fragments of molinate and molinate degradation product (keto-molinate isomers)

Molinate	Keto-molinate isomers	Functional group lost
$M^+ = 187$ 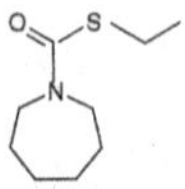	$M^+ = 201$	
$m/z = 126$ 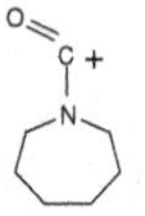	$m/z = 140$ 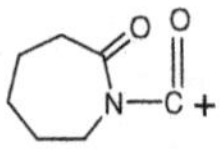	- [S-CH_2-CH_3]
$m/z = 98$ 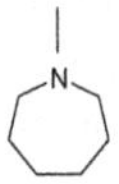	$m/z = 112$ 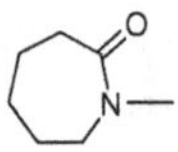	- [C=O]
$m/z = 83$ (C_6H_{11})	No ion	- [N-H]
No ion	$m/z = 84$ 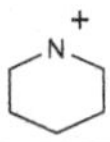	- [C=O]
No ion	$m/z = 69$ (C_5H_9)	- [N-H]

4.3.1. Effect of ZVI at Fixed Hydrogen Peroxide Concentration

Investigations of the effect of ZVI concentration on molinate degradation in the presence of hydrogen peroxide (H_2O_2) were undertaken. As shown in the Figure 4.22, addition of 50 mM H_2O_2 to ZVI of different concentrations significantly enhanced the degradation rate of molinate. Approximately 40% of the molinate is removed over 3 h in the presence of 50 mM H_2O_2, and the rate is further enhanced on addition of ZVI. Depending on the initial ZVI doses and the ratios, there is an initial sharp fall in concentration followed by slower degradation that results in the removal of between 70 and 100% of molinate over 3 h (Figure 4.22). The biphasic shape of the degradation curves suggests that Fe(II) is rapidly consumed on addition of H_2O_2, but there is slower continued degradation following the initial high activity.

4.3.2. Effect of Hydrogen Peroxide at Constant ZVI

The effect of low levels of peroxide at constant ZVI was examined, and the results are given in Figure 4.23. The results indicate that while low levels of hydrogen peroxide (e.g. 0.033 mM and 0.33 mM H_2O_2) have little ability to degrade molinate, its use in combination with 1.8 mM ZVI significantly enhances the rate of degradation. The degradation results indicate that there is no additional enhancement in degradation if only 33 μM of H_2O_2 is added to 1.8 mM of ZVI. This level

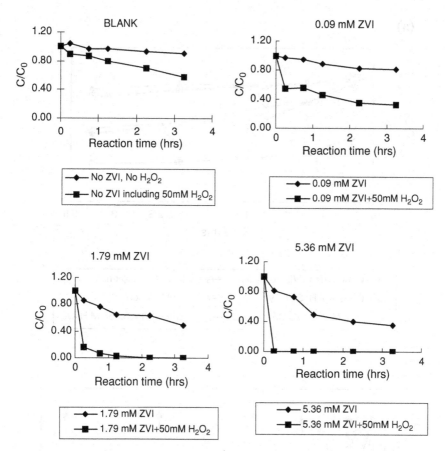

FIGURE 4.22. The effect of ZVI dose at fixed hydrogen peroxide. Condition: unbuffered ($pH_0 \sim 6.4$), 5 ppb molinate.

of H_2O_2 can therefore be considered a lower limit for enhanced degradation and may be similar to the transient levels of H_2O_2 initially generated on addition of ZVI to the molinate solution.

Molinate degradation efficiency was increased with increasing H_2O_2 dose. The effect of ferrous iron only in the absence of ZVI was also investigated to assess whether there is any effect of the oxygenation of Fe(II) to Fe(III) on molinate degradation. When ferrous iron alone is added to molinate, no degradation is observed (Figure 4.23).

4.3.3. Degradation By-products by Combined ZVI and H_2O_2

Since similar molinate degradation behavior was observed with both ZVI and coupled ZVI/H_2O_2, extraction experiments were conducted to examine whether there was any difference in the by-products generated for both processes.

(a)

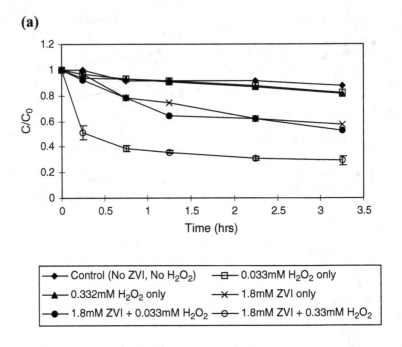

(b)

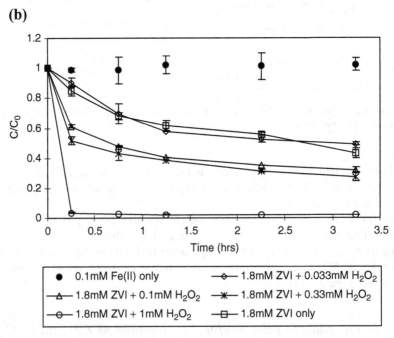

FIGURE 4.23. The effect of peroxide at constant ZVI: molinate (100 ppb) degradation in the presence of Fe^0 and H_2O_2 ($pH_0 = 6.4 \pm 0.3$).

(a)

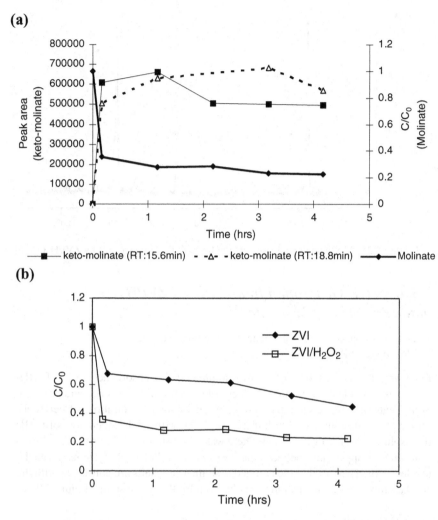

FIGURE 4.24. (a) Disappearance of molinate and production of by-product by ZVI/H_2O_2. (b) Molinate degradation by ZVI (19.6 mM) and coupled ZVI (19.6 mM) and H_2O_2 (20 mM).

By-product experiments were conducted using 10 ppm molinate, 19.6 mM ZVI, and 20 mM of H_2O_2 for the coupled system. After a 15-min reaction for the ZVI-only system and 10 min for the coupled system, all ZVI particles were extracted using 2 mL hexane, followed by sonication for 20 min, centrifugation, and direct injection into the GC/MS. The extent of molinate degradation was 33% for the ZVI-only system and 64% for the coupled ZVI/H_2O_2 system (Figure 4.24). The major by-products in both cases were identified to be keto-molinate isomers (Figure 4.25).

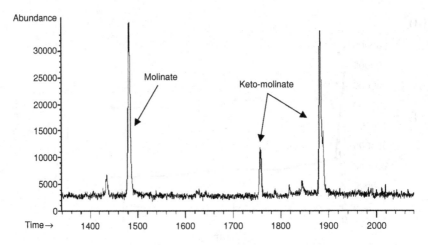

FIGURE 4.25. Molinate degradation and the formation of by-products by ZVI/H$_2$O$_2$.

4.3.4. Fe(II) Generation from Coupled ZVI/H$_2$O$_2$ in the Presence of Molinate

4.3.4.1. Comparison of Fe(II) Generation in the Absence and Presence of Peroxide

Dissolved Fe(II) was quantified by monitoring the absorbance of an Fe(II)-bipyridine complex at 522 nm (Voelker, 1994) both in the absence and presence of molinate. As can be seen in Figure 4.26, the initial concentrations of Fe(II) observed in solution increases with increasing ZVI, with a subsequent loss of Fe(II) from solution occurring over time. There was no substantial difference in Fe^{2+} evolution and disappearance between the ZVI-only and ZVI/H$_2$O$_2$ systems, given the low levels of H$_2$O$_2$ used in the coupled system. The results are consistent with the similar rates of molinate degradation observed for the two systems (Figure 4.23b).

4.3.4.2. Fe^{2+} Release in the Absence of Molinate (Online Continuous Measurement)

Both adsorbed and dissolved Fe^{2+} generated during the ZVI/H$_2$O$_2$ process was measured using the continuous analytical system. In the ZVI/H$_2$O$_2$ (0.33 mM H$_2$O$_2$) case, both the total and dissolved Fe^{2+} fractions decrease after 30 min (Figure 4.27). At higher H$_2$O$_2$ concentrations (3.3 mM), the total Fe^{2+} is considerably lower and very little dissolved Fe^{2+} is observed.

4.4. Molinate Degradation Using Fenton's Reagent

Fenton's reagent is a mixture of ferrous iron and hydrogen peroxide that leads to the generation of hydroxyl radical (OH$^{\bullet}$) through the iron-catalyzed decomposition

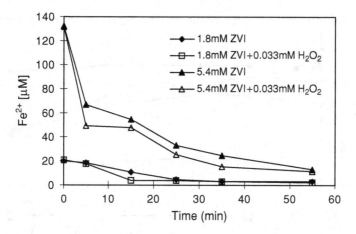

FIGURE 4.26. Dissolved Fe(II) variations during molinate degradation by H_2O_2/ZVI in pure water ($pH_0 \sim 6.4$) and molinate (100 ppb).

of hydrogen peroxide in acidic solution:

$$H_2O_2 + Fe^{II} \rightarrow OH^\bullet + OH^- + Fe^{III} \tag{4.6}$$
$$k = 76 \text{ M}^{-1}\text{s}^{-1}(\text{Mill and Haag, 1989})$$

pHs of <3.5 are optimal as (or because) this ensures high solubility of iron as well as a higher reducing potential. At higher pHs, hydrogen peroxide decomposes to water and oxygen resulting in excessive consumption of H_2O_2:

$$H_2O_2 \rightarrow {}^1\!/_2O_2 + H_2O \tag{4.7}$$

As can be seen in Figure 4.28a, the extent of molinate degradation is dependent on

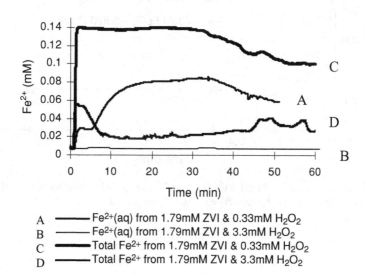

A	—— Fe^{2+}(aq) from 1.79mM ZVI & 0.33mM H_2O_2
B	—— Fe^{2+}(aq) from 1.79mM ZVI & 3.3mM H_2O_2
C	—— Total Fe^{2+} from 1.79mM ZVI & 0.33mM H_2O_2
D	—— Total Fe^{2+} from 1.79mM ZVI & 3.3mM H_2O_2

FIGURE 4.27. Fe^{2+} release from ZVI in presence of H_2O_2 ($pH_0 \sim 6.4$).

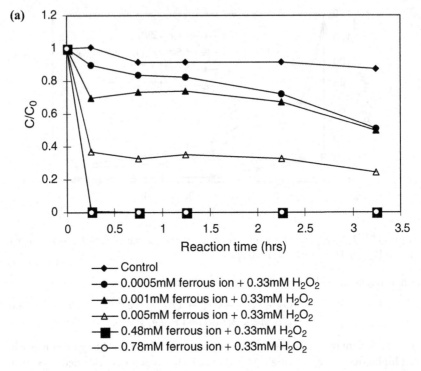

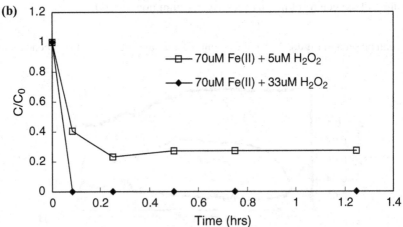

FIGURE 4.28. Degradation kinetics using Fenton's reagent under different pHs for the removal of molinate (100 ppb): (a) $pH_0 = 6.4 \pm 0.3$; (b) pH = 4.

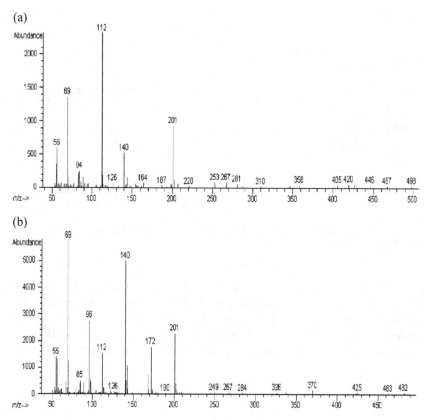

FIGURE 4.29. Mass spectrum of the GC peaks that appear in Fenton process at 17.6 min (a) and 18.9 min (b).

ferrous iron availability when H_2O_2 is in excess. The removal kinetics of molinate is initiated by rapid decay followed by much slower degradation. The slow removal is likely to be due to the slow regeneration of Fe(II) from the oxidized Fe(III). When Fe(II) rather than H_2O_2 is in excess (Figure 4.28b), the reaction ceases after 15 min as there is insufficient H_2O_2 for further degradation.

4.4.1. Degradation By-products of Molinate Using Fenton's Reagents

As can be seen in the Figures 4.29, 4.30, and 4.31, the keto-molinate isomers, which were also found in ZVI and ZVI/H_2O_2, were identified as by-products during degradation using Fenton's reagents. The finding strongly suggests that when using ZVI or the coupled H_2O_2/ZVI system, molinate is oxidized by radical species through a Fenton-like reaction. However, one of the isomers (exhibiting a retention time of 18.8 min) was not seen in the standard Fenton treatment. The retention

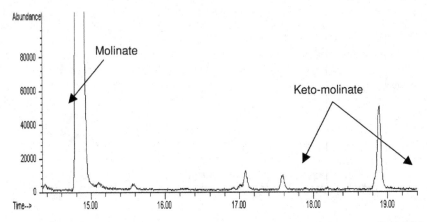

FIGURE 4.30. Molinate and its by-products after 3.25 h by dark Fenton process (0.11 mM Fe^{2+}/20 mM H_2O_2).

times of the characteristic ions of the intermediates are summarized in Table 4.4. The two GC chromatograms (Figures 4.30 and 4.31) indicate that in the Fenton process, the concentrations of the by-products are lower than in the ZVI-mediated processes, suggesting that the by-products are susceptible to oxidation. Hydroxyl radical reactions with the thiocarbamates (e.g., EPTC, vernolate, butylate, cycloate,

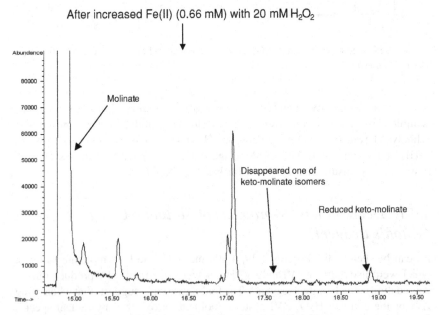

FIGURE 4.31. Molinate and its by-products after 15 min by dark Fenton process: increased Fe^{2+} (0.66 mM) with 20 mM H_2O_2.

molinate) are expected to proceed by H-atom abstraction from C—H bonds of the alkyl substituent groups or by addition to the N or S atoms of the NC(O)S groups. OH, CHO, and C=O group containing thiocarbamate derivatives are the primary intermediates of thiocarbamates (Vidal et al., 1999), and this is consistent with the observation of keto-molinate isomers during the degradation of molinate with ZVI, ZVI/H$_2$O$_2$, and dark Fenton process.

4.5. Comparison of ZVI, Coupled ZVI/H$_2$O$_2$ and Fenton's Process at High pH

Molinate degradation in a carbonate-buffered solution was investigated for the H$_2$O$_2$/ZVI system and dark Fenton process and compared with the ZVI-only case. In all cases the degradation rate of a solution containing 2 mM NaHCO$_3$ (pH 8.1) was considerably slower (Figure 4.32a–c) than observed when starting with a lower intial pH (Figure 4.6), especially for the H$_2$O$_2$/ZVIand dark Fenton processes. For these two processes, molinate degradation ceased completely after 1 h. In contrast, continuous ZVI degradation was observed during the ZVI process, especially for higher initial ZVI concentrations, although the degradation rate began to plateau after 2 h. Increasing concentrations of the ZVI, however, did not significantly improve the degradation rate (Figure 4.32c), unlike at lower pHs (Figure 4.6). The reduction in the degradation of molinate in the bicarbonate-buffered system is most likely due to the formation of a deactivating oxide layer on the surface of the particles, competitive adsorption from excess carbonate species for surface sites, faster oxidation of ferrous to ferric iron at higher pHs, and bicarbonate scavenging of hydroxyl radicals.

4.6. XRD and XPS Analysis

Since the morphology of a solid controls its dissolution and growth and the atomic structure on various crystal faces determines preference for adsorption of contaminants (Stipp et al., 2002), X-ray photoelectron spectroscopy (XPS) analysis would be a useful tool for determining the surface chemical structure information, and so would be X-ray diffraction (XRD) analysis for determining the bulk composition of the particles.

4.6.1. Results of XRD Analysis

4.6.1.1. Results of XRD Analysis in the Presence of Molinate

In general, corrosion products of iron consist of ferric oxyhydroxides (α-, β- and γ-FeOOH), magnetite (Fe$_3$O$_4$), and amorphous iron oxide (Ishikawa et al., 1998).Corrosion products formed during molinate degradation in pure water and in bicarbonate-buffered solution were characterized in order to assess their likely

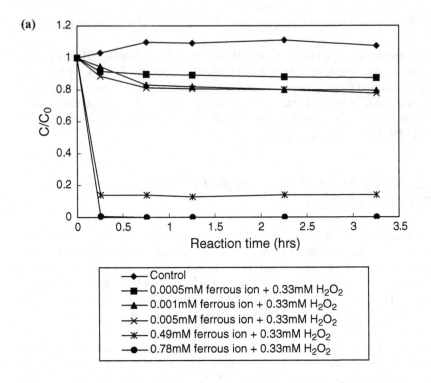

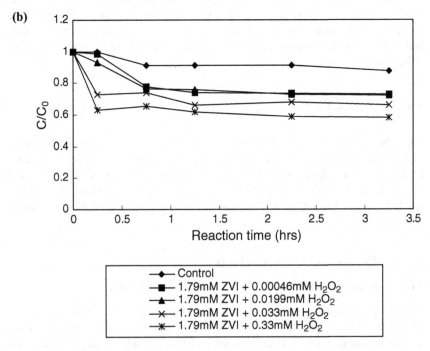

FIGURE 4.32. Bicarbonate effect (pH 8) on the degradation of molinate (100 ppb) by Fenton (a), H_2O_2/ZVI (b), and ZVI (c).

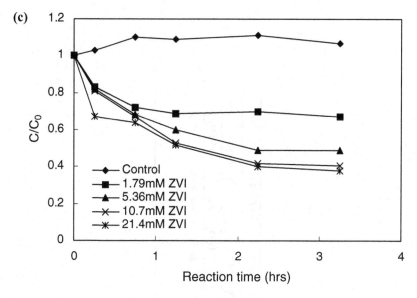

F IGURE 4.32. (*Continued*)

impact on molinate degradation kinetics. Samples were collected after a 3-h reaction time, and the results are summarized in Table 4.2 (See Appendix A for XRD). As can be seen, corrosion products collected from H_2O_2/ZVI and ZVI in pure water were identified as magnetite (Fe_3O_4), a black or brownish-red iron oxide, and maghemite-C or maghemite-Q (γ-$Fe_2^{III}O_3$). In comparison, the XRD peaks from the bicarbonate-buffered samples were consistent with those from ferric hydroxide, $Fe(OH)_3$.

4.6.1.2. XRD Results of ZVI Particles in the Absence of Molinate

In the absence of molinate, all four ZVI samples collected from each of the different reaction systems (after 3 h) were identified as a combination of iron and maghemite-C (γ-$Fe_2^{III}O_3$). It is known that the first corrosion product of Fe^0 under anaerobic conditions is $Fe(OH)_2$, which may be further oxidized to magnetite Fe_3O_4. Prior to the formation of magnetite, mixed-valent Fe(II)+Fe(III) salts, known as green rusts, may form under neutral pH conditions. The oxidation of mixed-valent salts commonly leads to the formation of maghemite (γ-Fe_2O_3) (Ishikawa et al., 1998). Charlet et al. (1998) revealed by detailed analysis that the corrosion products adhering to iron after oxidation by dissolved oxygen are hematite (α-Fe_2O_3), nonstoichiometric magnetite (Fe_3O_4), and lepidocrocite (γ-FeOOH). Odziemkowski (2000) showed that the electron-conducting magnetite layer could be oxidized to maghemite (γ-Fe_2O_3) at near-neutral pH, thereby stopping electron transfer from the Fe^0 core and halting the redox reaction. However, when aqueous Fe^{2+} ions were adsorbed and incorporated into the maghemite lattice, maghemite could be locally

TABLE 4.2. XRD results from four different samples

XRD	In the presence of molinate				In the absence of molinate			
	Pure water		Bicarbonate		Pure water		Bicarbonate	
	ZVI	ZVI/H_2O_2	ZVI	ZVI/H_2O_2	ZVI	ZVI/H_2O_2	ZVI	ZVI/H_2O_2
Products	Fe_3O_4 and γ-$Fe_2^{III}O_3$	Fe_3O_4 and γ-$Fe_2^{III}O_3$	$Fe(OH)_3$	$Fe(OH)_3$	γ-$Fe_2^{III}O_3$	γ-$Fe_2^{III}O_3$	γ-$Fe_2^{III}O_3$	γ-$Fe_2^{III}O_3$

Note: The ZVI concentration was 1.8 mM and that of H_2O_2 was 0.33 mM in all cases.

TABLE 4.3. XPS results (in the presence of molinate) and atomic concentration [AT]%.

| Samples | 1.79 mM ZVI/ H_2O_2 | | 1.79 mM ZVI | |
Peak	[AT]%	Identity	[AT]%	Identity
O 1s	5.6	Organic	4.5	Organic
O 1s	22.4	Fe hydroxide	22.4	Fe hydroxide
O 1s	30.0	Fe oxide	29.6	Fe oxide
C 1s	2.1	O=C—OH	1.6	O=C—OH
C 1s	0.6	O=C	0.6	O=C
C 1s	2.4	C—O	1.7	C—O
C 1s	18.0	C—C / C—H	20.3	C—C / C—H
Fe 3p	18.9	Fe(III)	19.4	Fe(III)

converted to magnetite, enhancing the conductivity of the oxide layer and allowing transfer of electrons from Fe^0 to the contaminant at the oxide/water interface.

4.6.2. XPS Results

The surfaces of the ZVI particles were analyzed using XPS. All the ZVI particles were collected after a 3.25-h reaction time and dried under nitrogen. Since the drying temperature under nitrogen may result in rapid surface oxidation, drying conditions were changed by using liquid nitrogen to minimize rapid oxidation in air, followed by drying 20 h using a vacuum drier. The results are reported in Table 4.3 and indicate that there is little difference on the particle surface regardless of whether degradation occurs using ZVI or ZVI/H_2O_2. The majority of the surface after reaction with molinate consists of iron oxide forms that have been oxidized to Fe(III). The high C—C/C—H ratio is indicative of hydrocarbon surface contamination from samples exposed to air. In the high vacuum used for the analysis, molinate presumably volatilizes as no N or S is seen in the samples.

The observations from XPS analysis are consistent with those of Devlin et al. (1998).

4.7. Discussion

4.7.1. Evidence of Oxidation Pathway

4.7.1.1. Keto-molinate By-products

The nature of the reaction end product (the keto-molinate isomer) indicates that ZVI is initiating an oxidation rather than a reduction process. Indeed, the keto-molinate isomer has previously been reported in molinate degradation studies using light and TiO_2 semiconductor particles resulting from hydrogen abstraction as a result of hydroxyl radicals preferentially attacking the N-alkyl chain (Konstantinou et al., 2001). Although the reduction of oxygen by Fe^0 is generally envisaged as a four-electron step with water as the major product (Equation (4.8)), a two-electron

TABLE 4.4. GC-MS-EI retention times (RT) and spectral characteristics of molinate identified by-products in both ZVI and ZVI/H$_2$O$_2$ and standard Fenton processes

By-products	Retention time (min)	Characteristic ions, m/z (abundance %)
Keto-molinate isomer (H$_2$O$_2$/ZVI, dark Fenton, ZVI)	17.57	112 (100), 69 (44), 140 (25), 201(45), 56(17)
Keto-molinate isomer (H$_2$O$_2$/ZVI, ZVI)	18.82	112 (100), 69 (22), 140 (65), 201(35), 56(17)
Unknown (in ZVI)	18.435	128 (100), 62 (37), 85 (32), 57 (30)
Unknown (in H$_2$O$_2$/ZVI)	18.435	128 (100), 57 (88), 89 (32)

reduction of oxygen to hydrogen peroxide would also appear possible (Equation (4.9)).

$$O_2 + 2Fe^0 + 4H^+ \leftrightarrow 2Fe^{2+} + 2H_2O \qquad (4.8)$$

$$O_2 + Fe^0 + 2H^+ \leftrightarrow Fe^{2+} + H_2O_2 \qquad (4.9)$$

Hydroxyl radicals (or possibly ferryl species (Bossmann et al., 1998; Kremer, 1999)) may well be generated in these systems as a result of Fenton's reagent processes arising from the presence of both ferrous iron and hydrogen peroxide (Equation (4.10))

$$Fe^{2+} + H_2O_2 \rightarrow Fe^{III}OH^{2+} + OH^\bullet \qquad (4.10)$$

Whether this process occurs principally in solution or at the surface of the ZVI particles is unclear at this stage, but it is highly likely that many of the key steps occur at the surface of the ZVI particles. The proposed oxidation mechanism was further confirmed with the following results.

4.7.1.2. Same By-products for Different Oxidation Processes

The same by-products (keto-molinate isomers) were found in all three processes (ZVI, ZVI/H$_2$O$_2$, standard Fenton) (Table 4.4). The result indicates that the attack of $^\bullet$OH represents the major initiation step in the degradation pathway of molinate. It has been suggested that $^\bullet$OH can react via hydrogen-atom abstraction, electron transfer, or addition to the N atom, leading to carbon-centered and (or) nitrogen-centered radicals (Das et al., 1987; Hayon et al., 1970; Rao and Hayon, 1975; Simic et al., 1971). In this reaction, the hydroxyl radical reacts via the addition to the N atom, producing a carbon-centered radical; in the presence of dissolved oxygen, the R$^\bullet$ radical leads to the production of ROO$^\bullet$; and subsequent reactions lead to the end product, keto-molinate isomers. Addition of butanol, a hydroxyl radical scavenger, prevented the degradation of molinate (Figure 4.19).

Under anaerobic conditions, achieved by bubbling nitrogen gas through a reactor containing 10 ppm molinate and ZVI, no degradation or keto-molinate isomers were found (Figures 4.33 and 4.36). This suggests that the presence of oxygen plays an important role in molinate degradation.

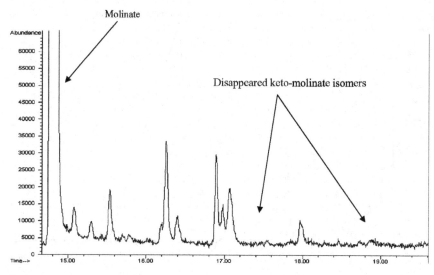

FIGURE 4.33. GC chromatogram: keto-molinate peaks disappeared in the absence of oxygen.

4.7.1.3. Effect of Catalase

The addition of catalase, which induces the rapid breakdown of hydrogen peroxide, prevented the formation of keto-molinate isomers (Figure 4.34) and degradation of molinate (Figure 4.18). This suggests that the presence of hydrogen peroxide, an intermediate product formed from the disproportionation of superoxide, is critical for degradation. The catalase results support the hypothesis that ferrous iron formed from iron corrosion reacts with hydrogen peroxide, resulting in the formation of hydroxyl radicals, the primary oxidant in the system.

4.7.1.4. Effect of SOD

Addition of the enzyme superoxide dismutase (SOD), which catalyzes the disproportionation of the superoxide radical $O_2^{\bullet-}$ to H_2O_2 (Fee and DiCorleto, 1973), accelerates the formation of hydrogen peroxide and leads to a slight improvement in the removal rate (Figure 4.36). Addition of SOD also resulted in the formation of keto-molinate isomers, as shown in Figure 4.35.

4.7.2. Reaction Mechanism

All these experimental results provide confirmative evidence for the proposed oxidation mechanism. For example, electrons produced on the oxidation of Fe^0 are likely to react with molecular O_2 adsorbed to the Fe^0 surface, possibly reducing

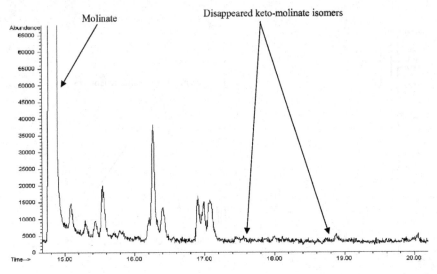

FIGURE 4.34. GC chromatogram: keto-molinate peaks disappeared in the presence of catalase/O_2.

it to the superoxide radical anion $O_2^{\bullet-}$ (or to the hydroperoxy radical $HO_2^{\bullet-}$ at pH < 4.8) as shown below.

$$Fe^0 \rightarrow Fe^{2+} + 2e^- \tag{4.11}$$

$$(O_2)_{ads} + e^- \rightarrow O_2^{\bullet-} \quad or \quad HO_2^{\bullet} \tag{4.12}$$

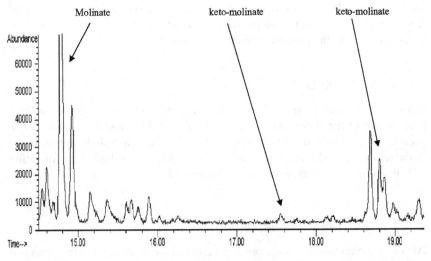

FIGURE 4.35. GC chromatogram: keto-molinate peaks appeared in the presence of SOD/O_2.

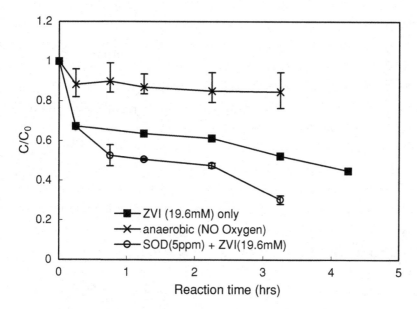

FIGURE 4.36. Effect of presence/absence of oxygen and presence/absence of superoxide dismutase (SOD). *Note*: 5μg/mL of SOD was added over 2.75 h.

Disproportionation of these free radicals is likely to occur relatively rapidly at low pH ($k_{12a} = 9.7 \times 10^7$ M^{-1}s^{-1} (Bielski et al., 1985)) (slower at higher pH; at pH 8, $k_{4.13b} = 6.1 \times 10^4$ M^{-1}s^{-1}), resulting in the formation of hydrogen peroxide at or near the ZVI surface (Equation (4.13)):

$$HO_2^\bullet + HO_2^\bullet \rightarrow H_2O_2 + O_2 \tag{4.13a}$$

$$HO_2^\bullet + O_2^{\bullet-} + H_2O \rightarrow H_2O_2 + O_2 + OH^- \tag{4.13b}$$

The hydrogen peroxide generated in this manner may react with the ferrous iron released during the ZVI corrosion process to produce highly reactive hydroxyl radicals.

Molinate is degraded by the attack of hydroxyl radical and then reacts with oxygen to form an organoperoxy radical (MolOO$^\bullet$) before forming the reaction end product, keto-molinate isomers, i.e.,

$$\text{Molinate} + \bullet OH \rightarrow \text{Mol} + H_2O \tag{4.14}$$

$$\text{Mol} + O_2 \rightarrow \text{MolOO}^\bullet \tag{4.15}$$

$$\text{MolOO}^\bullet \rightarrow \text{keto-molinate isomers} \tag{4.16}$$

The presence and absence of keto-molinate by-products under different reaction conditions strongly suggests that ZVI in the presence of O_2 is oxidatively rather than reductively degrading molinate. The rapid consumption of DO and

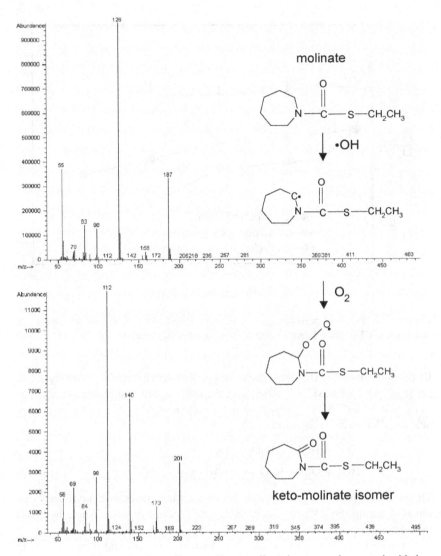

FIGURE 4.37. MS fragments for molinate and keto-molinate isomers and proposed oxidation pathway (Joo et al., 2004).

production of hydrogen peroxide, the enhancement in molinate degradation when supersaturating the reaction solution with O_2, and the absence of any appreciable degradation in the absence of O_2 further support an oxidative mechanism. There are still many unknown facts, however; in particular, whether the oxidation occurs on the surface of the particles or in the bulk solution. These issues are further investigated in the next chapter using specific model organics.

4.7.3. Kinetics of Fe(II) and H_2O_2 Generation

The proposed pathway would appear to qualitatively account for the observed kinetics of Fe(II) and H_2O_2. For example, the slow decrease in ferrous iron concentrations in pH 4 solutions most likely represents a balance between Fe^{2+} release from the oxidizing Fe^0 surface and subsequent oxidation of this ferrous iron by hydrogen peroxide to ferric species. Hydrogen peroxide concentrations reach a steady state at pH 4 but only after initial rapid reaction with released ferrous iron. It is presumably because of this initial consumption of hydrogen peroxide by ferrous iron that a lag in accumulation of H_2O_2, particularly at the higher ZVI loadings, is observed. The kinetic rate can be affected by concentrations of Fe^{2+} as well as the reactive ZVI surface area available. The significant dependence of degradation rate on ZVI concentrations is mainly due to a greater accumulation of Fe(II) (aq) in the suspension. As the reaction proceeds, the concentration of Fe(II) (aq) decreases with this effect more pronounced in the bicarbonate system, where iron corrosion is inhibited by the formation of an oxide layer on the particles.

At higher pH (8.1), oxidation of Fe^{2+} by oxygen will proceed more rapidly resulting in very little accumulation of Fe^{2+} in solution. In addition, the tendency of Fe^{2+} to remain adsorbed to the surface of either Fe^0 or to iron oxyhydroxide particles (formed on hydrolysis of ferric species present as a result of Fe^{2+} oxidation) would also be expected to lower the measurable concentration of Fe^{2+} in solution.

4.7.4. Overview of the ZVI-Mediated Oxidative Technology

The results of the molinate study indicate that nanoscale ZVI particles are capable of oxidizing and thus degrading certain organic contaminants in the presence of oxygen. The reaction is abiotic and pH dependent (slower at higher pH levels). The formation of iron oxyhydroxide particles at alkaline pH might also be expected to lower the reactivity of the ZVI as a result of surface accumulation; however, relatively similar rates of molinate degradation are observed at pH 4 and 8.1, at least over the first hour or so. In such high carbonate content (2 mM $NaHCO_3$) solutions used in this study, it is also possible that siderite ($FeCO_3$(s)) could initially form (thereby lowering the ferrous iron activity). Its occurrence would only be transitory, however, in view of the continual oxidation of Fe(II) to Fe(III) species.

Passivation of the Fe^0 metal surface by these precipitates is one of the most likely limitations to effective long-term performance of permeable reactive barriers. In a traditional equilibrium analysis of corrosion potentials (Pourbaix, 1966), conditions that favor the formation of $FeCO_3$(s), $Fe(OH)_2$(s), and $Fe(OH)_3$(s) should passivate Fe^0, whereas sustained corrosion is expected only below pH 6 where the oxidation product is aqueous Fe^{2+}.

The continued degradation of molinate for 2 h or more at pH 8.1 is confirmation of the continuing release of ferrous iron, formation of reduced oxygen species, and the subsequent generation of strongly oxidizing substrates (such as the hydroxyl radical, or possibly ferryl species). Any ferrous iron initially present in the reaction medium (derived from the original ZVI stock) would expect to be oxidized in a few minutes with the resultant cessation of oxidant generation.

While the eventual reduction in the extent of molinate degradation at higher pH may reflect the effect of surface coatings, it is intriguing to observe such a high degree of degradation as most Fenton reagent mediated processes are relatively ineffective at higher pH. The difference in this case is the continuing generation of fresh reactants (Fe^{2+} and H_2O_2), with the effectiveness of the process at high pH limited by Fe^0 availability.

The slow release/formation of the key reactants and their continuing effectiveness at degrading the contaminant of interest even at high pH suggests a range of possible applications for ZVI-mediated oxidative processes ranging from in situ degradation of contaminants in oxic groundwaters, degradation of contaminants on surface sediments (or, indeed, surfaces in general), and degradation of contaminants in water treatment by periodic placement of nanosized ZVI particles on sand or anthracite components of deep bed filters. Such processes could be initiated by the simple emplacement of ZVI particles in oxic environments, with the ensuing degradation of contaminants achieved without the need for continuous dosing of wet chemicals.

4.8. Conclusion

Degradation of the carbothiolate herbicide, molinate, has been investigated in oxic solutions containing nanoscale ZVI particles and found to be effectively degraded by an oxidative pathway. Both ferrous iron and superoxide (or, at pH $<$ 4.8, hydroperoxy) radicals appear to be generated on corrosion of the ZVI, with resultant production of strongly oxidizing entities capable of degrading the trace contaminant. While researchers have found the reductive degradation pathways in anoxic Fe^0/H_2O systems, which involve either direct electron transfer from ZVI at the surface of the iron metal or reaction with dissolved Fe^{2+} or H_2, molinate degradation studies under aerobic conditions showed the presence of oxidized degradation by-products. These results suggest that molinate degradation occurs through an oxidative mechanism even though the ZVI process is normally considered highly reductive. At the trace levels of organics being used in this study, oxygen is expected to be the primary electron acceptor and superoxide the primary radical intermediate.

Investigation of degradation pathways indicates the important role of radical species in treating molinate by ZVI. Studies on Fe^{2+} and Fe^{3+} generation during ZVI and ZVI/ H_2O_2 treatment suggest that Fe^{2+} availability and surface dissolution kinetics may play a role in the degradation of molinate. The combined H_2O_2/ZVI system was found to be highly effective and reduces the quantity of ZVI necessary for rapid removal of molinate for comparatively low H_2O_2 doses. However, in the presence of bicarbonate the removal of molinate was retarded, possibly because of the rapid oxidation of Fe(II) to Fe(III) and the inhibiting role of carbonates toward free-radical reactions. Preventing the formation of surface oxide films at high pH and maintaining ferrous iron availability would improve the applicability for treatment of pesticides in alkaline waters.

The ZVI-mediated oxidation process is promising, although further research is required to understand the reaction mechanism, the particles lifetime, and factors that affect ZVI reactivity. Thus, the following chapter investigates the particles in greater detail by using specific organic probes to assess the oxidation mechanism and determine the oxidizing capacity of ZVI.

5

Molecular Oxygen Activation by Fe$^{II/III}$EDTA as a Form of Green Oxidation Chemistry*

5.1. Oxygen Activation

There is a global need for new environmentally clean, industrial technologies, or so-called green chemistries (Allen et al., 2002; Lancaster, 2002; Tundo et al., 2000). Much attention has been focused on the goal to develop and refine a green oxidation system capable of degrading key priority organic pollutants (or xenobiotics). Many of these species that are persistent and bioaccumulative include dioxins, furans, chlorinated phenols, and many pesticides; once released into the environment, they may linger there for decades (EPA, 2001). An ideal oxidation scheme would involve complete mineralization or deep oxidation of a broad spectrum of these xenobiotics. Coupled with green chemistry concepts, it is best if the system were to do this under mild conditions using abundant nontoxic reagents. A newly developed system fulfills these criteria. Deep oxidation of xenobiotics was found to occur in the presence of zero-valent iron (ZVI), ethylenetetraaminediacetic acid (EDTA), air (ZEA) under aqueous conditions (Noradoun and Cheng, 2005; Noradoun et al., 2003; Noradoun et al., 2005). A main feature of this system is its ability to create reactive oxygen species in situ from O_2(aq) using simple, easily stored, and safe reagents under mild reaction conditions. ZVI is an attractive component for a green chemistry scheme because it is inexpensive and abundant, as well as essentially nontoxic along with its spent counterpart, Fe^{2+} ions.

To date, this is the only known system to obtain near-complete mineralization of organic pollutants through nonbiological oxygen activation under room temperature and 1 atm pressure conditions. Oxygen activation by nonbiological systems has been highly desired as an effective, low-cost, and green method for inorganic and organic syntheses as well as detoxification of pollutants. For these reasons, there has been a considerable amount of effort concentrated in this field of research. Life on earth depends on molecular oxygen, but an overlooked aspect is that its kinetic stability enables organic life forms to flourish without the deleterious effects of spontaneous oxidations. Dioxygen stability is due to its spin-triplet ground state,

* Contribution from Christina Noradoun and I. Francis Cheng, University of Idaho, Moscow, Idaho 83844-2343.

the two unpaired electrons in the highest occupied molecular orbitals (Sawyer, 1991; Simandi, 1992). Since organic and biological molecules have paired electrons and singlet ground states, their reactivity with oxygen is spin forbidden. This kinetic barrier can be overcome in one of the following three ways: first, by exciting the O_2 to one of its singlet states (usually done photochemically); second, by complexation with a paramagnetic metal ion, such as iron; or third, by a radical mechanism that is capable of circumventing the spin restriction (Sawyer, 1991). Activation of dioxygen under mild, reductive conditions is most widely known to occur naturally in heme-containing metalloenzyme systems such as cytochrome P450 enzymes and methane monooxygenase (Sharma et al., 2004). These enzymes utilize molecular oxygen to perform selective oxidation reactions. There are known examples of catalytic oxidation by dioxygen of all major types of saturated hydrocarbons (aliphatic, cycloaliphatic, and alkyaromatic) by the large family of cytochrome P450 enzymes (Simandi, 1992). With regard to the mechanism of these biological (nonheme and heme) processes, it has been debated whether, or under which conditions, the oxidation is mediated by Fe^{III}-hydroperoxo, $Fe^{IV}=O$ (ferryl), $Fe^V=O$ (perferryl), or diiron/dioxygen $Fe^{II}O—OFe^{III}$ complexes (Feig and Lippard, 1994; Feig et al., 1996). Also important are the partially reduced forms of O_2, the reactive oxygen species (ROS): $O_2^{\bullet-}$ (superoxide) and H_2O_2. The increased reactivity of the latter two species can be rationalized partially on the basis of the lowering of the bond order from 2 for O_2 to 1.5 and 1 for $O_2^{\bullet-}$ and H_2O_2, respectively.

While most research done on abiotic dioxygen activation has focused on iron porphyrins as biological mimics (Simandi, 1992), Barton and Doller (1992) have discovered a nonporphyrin iron, mixed-solvent catalyst scheme known as the "Gif system." Gif reactions are capable of catalyzing mono-oxgenation of carbon/hydrogen bonds to produce ketones and alcohols (Seibig and van Eldik, 1999).

$$RH + O_2 \, (2H^+, 2e^-, M^{n+}) \rightarrow ROH + H_2O \tag{5.1}$$

Gif chemistry has achieved distinction because of its regioselectivity. As a result, the Gif reaction can be viewed primarily as a synthetic method (Noradoun and Cheng, 2005). Furthermore, as a method for large scale environmental remediation, the Gif reaction would suffer from the drawbacks of expense and toxicity of its solvent system, pyridine (Stravropoulos et al., 2001).

Another strongly oxidizing system using iron salts and H_2O_2 has been known for more than 100 years as the Fenton reaction (Fenton, 1894). This reaction (Equation (5.2), where L indicates a suitable ligand) employs the hydroxyl radical (OH•) as a potent oxidizing agent capable of oxidizing most organics with diffusion-limited kinetics (Stadtman and Berlett, 1991).

$$Fe^{II}L + H_2O_2 \rightarrow Fe^{III}L + HO^- + OH\bullet \tag{5.2}$$

Other transition metals and peroxides can be used to create the OH•, and these systems are called "Fenton-like" reactions.

To date, several investigators have examined the possibility of using the Fenton and Fenton-like systems for the deep oxidation of organic pollutants, i.e. oxidation to inorganic carbon compounds (Augusti et al., 1998; McKinzi et al., 1999; Pratap

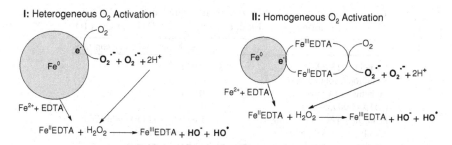

FIGURE 5.1. Proposed reaction scheme for ZEA oxidation system.

and Lemley,1998; Sedlak and Andren,1991; Watts et al., 1999; Watts et al., 2000). However, in most cases the specific procedure requires one or more of the following conditions: the use of acid pH (<4), photo-assistance in the near-UV, elevated temperatures (35–100°C) or high concentrations of H_2O_2 relative to the pollutant. Hydrogen peroxide is generally regarded as an environmentally "green" reagent, partly because it avoids the production of halogenated by-products associated with chlorine-based oxidations (Bogdal et al., 2004). However, H_2O_2 at high concentrations (typically 30% or greater) is hazardous to store, transport, and handle, and is also expensive. The ZEA reaction differs from other Fenton-type systems in that reactive oxygen species (i.e. H_2O_2) are produced in situ via oxygen activation by the $Fe^{II/III}EDTA$ complex (Figure 5.1).

Both schemes would use ZVI as a pool of reducing agent. The Fe^{2+} ions that are produced by the spontaneous corrosion of ZVI are complexed by EDTA forming $Fe^{II}EDTA$. Aqueous O_2 is either reduced by the Fe(0) (scheme I) or $Fe^{II}EDTA$ (scheme II) or by both routes to more reactive forms. These activated oxygen species may include $O_2^{\bullet-}$ and H_2O_2. Superoxide may dismutate to H_2O_2, either through metal-catalyzed or through noncatalyzed processes. A final reduction process, through the Fenton reaction, yields OH•. The process is proposed to redox cycle $Fe^{III}EDTA$ to $Fe^{II}EDTA$ as long as there is an excess of ZVI.

5.2. Xenobiotic Degradation by ZEA System

A typical demonstration system consisted of 2.5 g 20-mesh ZVI, 1 mM EDTA, and 1 mM xenobiotic in 50 mL of aqueous solution under aerobic conditions (Noradoun and Cheng, 2005; Noradoun et al., 2003; Noradoun et al., 2005). The mechanisms and products of the ZEA reaction are described below. Chlorinated phenols (Noradoun et al., 2003), organophosphorus compounds (Noradoun et al., 2005), and EDTA (Noradoun and Cheng, 2005) have been degraded using the ZEA system. All systems exhibit similar pseudo-first-order rate constants (Figure 5.2 and Table 5.1). Electrospray ionization/mass spectrometry (Agilent 1100) measurements of the post-reaction mixtures of 4-chlorophenol, malathion, and EDTA indicate low-molecular carboxylic acids as the major products (Noradoun and Cheng, 2005; Noradoun et al., 2003; Noradoun et al., 2005). In the absence of any xenobiotic EDTA degradation evident, the carbon balance is as presented

TABLE 5.1. Pseudo-first-order xenobiotic degradation rates for the ZEA O_2 activation system. Limit of detection (LOD) is for GC-FID

	Pseudo-first-order rate constants
Phenol (0.013 mM)	0.94 h^{-1}
4-chlorophenol (1.1 mM)	1.11 h^{-1}
Malathion (0.44 mM)	0.92 h^{-1}
EDTA (1.00 mM)	1.02 h^{-1}
Pentachlorophenol (0.61 mM)	Degraded to LOD within 70 h
Nitrobenzene (8.00 mM)	Degraded to LOD within 24 h

TABLE 5.2. Carbon balance of reaction products of the ZEA degradation of EDTA. Conditions: 1 mM EDTA, 2.5 g ZVI, air, reaction volume 50-mL, 6 h

	% C
$\leq C3^a + CO + CO_2$	35% ($\pm$ 5)
Iminodiacetic acid	28% ($\pm$ 3)
Oxalic acid	17% ($\pm$ 2)
Propionic acid	14% ($\pm$ 2)
EDTA	2% ($\pm$ 2)
Total	**96% ($\pm$ 5)**

^{a} Represents unidentified compounds of C3 or less excluding oxalic and propionic acid.

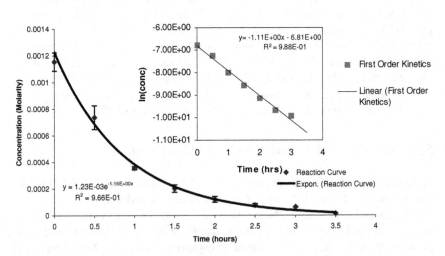

FIGURE 5.2. Kinetic data showing the first-order decay of 4-chlorophenol. The rate constant was found to be -1.11 h^{-1}. Each point on the graph represents the average peak area of three measurements taken on one individual sample mixture. The error bars show the standard deviation between the triplicate analyses.

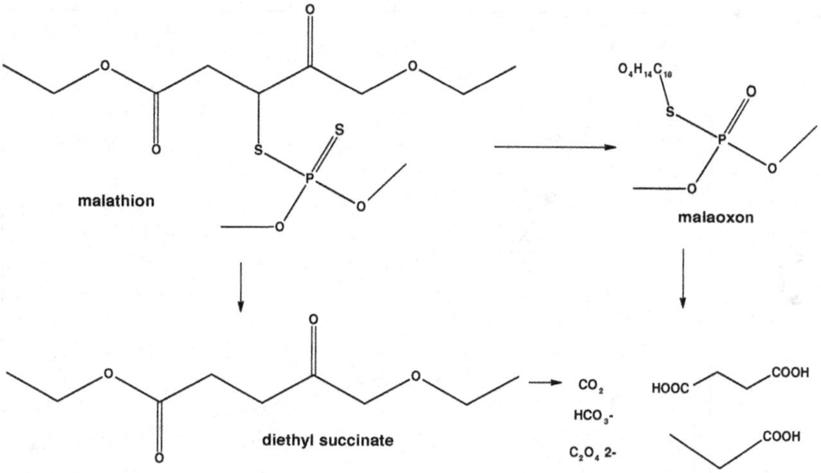

FIGURE 5.3. Malathion degradation scheme showing harsh oxidation of malathion and malaoxon to low-molecular-weight acids.

in Table 5.2. On the basis of mass spectroscopy results, the proposed degradation scheme for malathion is shown in Figure 5.3.

The presence of oxygen is crucial; the observed pseudo-first-order rate constants for EDTA removal are $k_{obs} = 1.02 \text{ h}^{-1}$ ($k_{obs,s} = 3.69 \pm 0.92 \text{ h}^{-1} \text{ m}^2$) and $k_{obs} = 0.04 \text{ h}^{-1}$ ($k_{obs,s} = 0.145 \pm 0.04 \text{ h}^{-1} \text{m}^2$) under air and under N_2 purge, respectively (Noradoun and Cheng, 2005). The ratio of [EDTA] to [$Fe^{II/III}$] is also important in controlling the reaction rate. Large excesses of EDTA were found to hinder the reaction (Figure 5.4); this will be examined in greater detail below.

5.3. Mechanism of Degradation

The mechanism by which the ZEA reaction proceeds is hypothesized to be through reactive oxygen species (ROS). Intermediates such as hydrogen peroxide are postulated to be continuously produced by the reduction of aqueous oxygen, which may take place either on the iron surface or in solution (Noradoun and Cheng, 2005; Seibig and van Eldik, 1997; Seibig and van Eldik, 1999). Many Fe^{II} complexes can react with O_2 to form the superoxide radical (Equation (5.4)) that leads to the production of H_2O_2 (Equation (5.5)) and eventually the Fenton reaction (Equation (5.6)) (Buettner et al., 1983; Bull et al., 1983; Rose and Waite, 2002).

$$Fe^0 \rightarrow Fe^{2+} + 2e^- \tag{5.3a}$$

$$Fe^{2+} + L \rightarrow Fe^{II} \tag{5.3b}$$

$$Fe^{II} + O_2 \rightarrow Fe^{III} + O_2^{\bullet-} \tag{5.4}$$

$$Fe^{II} + O_2^{\bullet-} + 2H^+ \rightarrow Fe^{III} + H_2O_2 \tag{5.5}$$

$$Fe^{II} + H_2O_2 \rightarrow Fe^{III} + OH^{\bullet} + OH^- \tag{5.6}$$

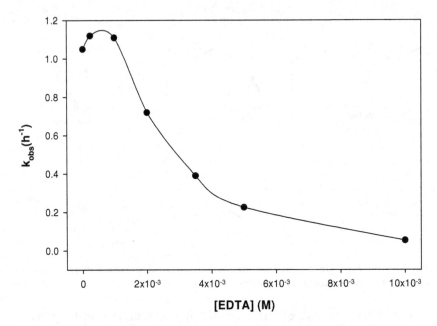

FIGURE 5.4. Pseudo-first-order rate constants for EDTA degradation affected by EDTA concentration. All systems mixed at 450 rpm, open to atmosphere, and unbuffered using 2.5 g ZVI, 1 mM EDTA, 50 mL total volume.

Free iron ions are precluded from taking part in the O_2 activation sequence shown in Equations (5.4)–(5.6).

Evidence for radical activity is supported by the suppression of EDTA degradation by a known hydroxyl radical scavenger, 1-butanol (Noradoun and Cheng, 2005; Psillakis et al., 2004; Vassilakis et al., 2004). Figure 5.5 illustrates that the ZEA reaction is effectively suppressed by the addition of 5 mM 1-butanol. Although not an exclusive proof that hydroxyl radical species are present in the reaction, it does provide evidence of reactive oxygen species being present in the reaction mixture. Further evidence is provided by using other assays (Cheng et al., 2003; Halliwell et al., 1987). Current research is under way measuring H_2O_2 as an intermediate in the ZEA reaction.

5.4. Rate-Determining Step

An examination of the rate-limiting step of the ZEA reaction was conducted by measuring the activation energy (E_a). An investigation of the degradation of EDTA at 3.2, 14.2, 22.2, 32.7, and 42.2°C, with all other parameters held constant, reveals an increase in the k_{obs} for EDTA degradation with an increase in temperature. The data when plotted using the Arrhenius equation $k = A \exp(-E_a/RT)$, and normalized with respect to the solubility of oxygen in water at the noted temperatures, established the linear behavior shown in Figure 5.6. An activation energy of 39.3 (± 0.6) kJ/mol was obtained in this temperature range.

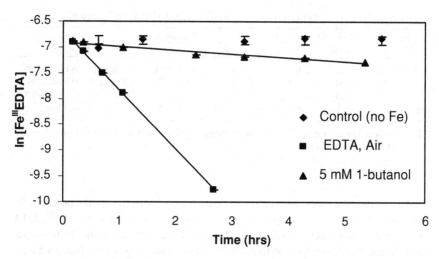

FIGURE 5.5. The suppression of EDTA degradation with the addition of 5 mM 1-butanol, pseudo-first order rate constants (■) k_{obs} = 1.22 h^{-1}, and with 5 mM 1-butanol (▲), k_{obs} = 0.08 h^{-1} (2.5 g ZVI g, 1.00 mM EDTA, open to air, 50 mL total volume).

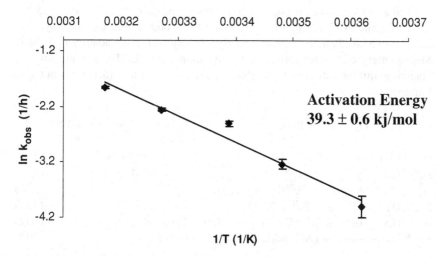

FIGURE 5.6. Arrhenius plot demonstrating temperature dependence of observed rate constants (k_{obs}) (2.5 g ZVI, 1 mM EDTA, 50 mL total volume).

TABLE 5.3. Activation energy values comparison of ZEA system to literature.

E_a (kJ/mol)	[FeEDTA] (mM)	[O_2] (mM)	pH	Temperature range	Reference
47.6 ($\pm$ 0.6)	2.5	0.125	5	13.5–49°C	Zang and van Eldik, 1990
33.9 ($\pm$ 1.4)	20	0.125	5	16–46°C	Zang and van Eldik, 1990
27.2 ($\pm$ 2.3)	100.0	–	7.5	–	Zang and van Eldik, 1990
39.3 ($\pm$ 0.6)	1.0	0.20–0.40[a]	5.5–6	3.2–42°C	Noradoun and Cheng, 2005

[a] Reaction vessel open to atmosphere; therefore oxygen solubility varied with temperature. This was accounted for in E_a calculations.

The activation energy measured can be compared with the results obtained by previous investigators measuring the rate of reduction of $Fe^{II}EDTA$ to $Fe^{III}EDTA$ by O_2(Wubs and Beenackers, 1993; Zang and van Eldik, 1990). Both studies indicate the rate-limiting step to be the O_2 reduction step represented in Equation (5.8) (Zang and van Eldik, 1990). The activation energy measured for this study is in good agreement (Table 5.3); the small discrepancies may be attributed to the different methods and conditions used in measuring the reaction kinetics.

The following reaction scheme shows oxygen activation to be the rate-limiting step (Equation (5.8)). Equations (5.7)–(5.13) have been theorized as the sequence of reactions important to the reduction of O_2 by $Fe^{II}EDTA$ (Zang and van Eldik, 1990). In our studies we hypothesize, in addition to Equations (5.7)–(5.11), Equations (5.12)–(5.13) that explain the production of the hydroxyl radical through the Fenton reaction and the consequential degradation of EDTA. Measurement of the activation energy by monitoring EDTA degradation and obtaining similar values indicate that Equations (5.12) and (5.13) are not rate limiting.

$$Fe^{II}EDTA + O_2 \rightarrow O_2Fe^{II}EDTA \qquad k_1 = 10^3 M^{-1}s^{-1} \ k_{-1} = 10^6 \quad (5.7)$$
$$(K_1 = k_1/k_{-1} = 10^{-3})*$$
$$O_2Fe^{II}EDTA \rightarrow Fe^{III}EDTA + O_2^- \qquad 7 \ k_2 = 10^2 \ M^{-1}s^{-1} \qquad (5.8)$$
$$Fe^{II}EDTA + O_2- \rightarrow O_2^{2-}\text{-}Fe^{III}EDTA \qquad k_3 = 10^6 \ M^{-1}s^{-1} \ [\dagger] \qquad (5.9)$$
$$O_2^{2-}\text{-}Fe^{III}EDTA + H^+ \rightarrow Fe^{III}EDTA + H_2O_2 \qquad k_4 = \text{fast} \ [\dagger] \qquad (5.10)$$
$$2Fe^{II}EDTA + H_2O_2 \rightarrow 2Fe^{III} + 2H_2O \qquad k_5 = 10^4 \ M^{-1}s^{-1} \ [\dagger] \qquad (5.11)$$
$$Fe^{II}EDTA + H_2O_2 \rightarrow Fe^{III}EDTA + OH^- + OH \qquad k_6 = 10^3 \ M^{-1}s^{-1} \ [\ddagger] \qquad (5.12)$$
$$Fe^{II/III}EDTA + OH \rightarrow LMW \ \text{acids} + CO_2 \qquad k_7 = \text{fast} \qquad (5.13)$$

Low-molecular-weight (LMW) acids = propionic, oxalic, and iminodiacetic acids.

[†] (Zang and van Eldik, 1990).
[‡] (Croft et al., 1992).

5.5. Iron Chelation and Chelate Geometry Influence Reactivity

Iron (II) is capable of reducing hydrogen peroxide, producing reactive hydroxyl radicals. The ability to do so depends on the chelation sphere determined by the chelating ligand (L). Variations in the ligand type or concentration may influence the thermodynamics and/or the kinetics of the electron transfer between iron, hydrogen peroxide, and/or oxygen, as well as the formation of the $Fe^{II}(L)$-O_2 adduct, determined earlier to be the rate-determining step in the ZEA reaction. EDTA was chosen as the primary chelating ligand for the ZEA reaction scheme because of its low cost and high solubility in aqueous solutions as well as the natural self-buffering nature of the disodium salt near pH 5.5.

The knowledge that irreversible oxidative damage can be done by the ROS present in the physiological systems has prompted a significant amount of research to elucidate the nature of biological ligands as either anti- or pro-oxidant in character (Blokhina et al., 1991; Cheng and Breen, 2000). Although completely anthropogenic, EDTA can be used as an analogue to proteins and amino acids found in biological systems (Englemann et al., 2003). When used in biological studies, it has been found that EDTA may exert both anti- and pro-oxidant behaviors (Englemann et al., 2003). Previous research with $Fe^{II/III}$EDTA has shown high ratios of [EDTA] to [Fe] (ratios of 2.5:1 and greater) to inhibit the Fenton reaction (Englemann et al., 2003). Antioxidant characteristics of the $Fe^{II/III}$EDTA complex are seen when EDTA is in large excess, and this is attributed to steric effects around the metal center, thus presenting a kinetic barrier (Englemann et al., 2003; Noradoun and Cheng, 2005). It can be concluded that not only the type of ligand is important, but also the [L] to [metal ion] ratio plays a fundamental role in how effective the ZEA system will be in promoting oxidation through reactive oxygen species.

Another important factor in the discussion of how effective a chelating ligand will be in promoting oxidation is denticity. The chelation sphere surrounding the iron center is strongly affected by the denticity of the ligand. Tetradentate ligands tetraamidomacrocyclic ligand (TAML) (Collins, 1994; Collins, 2002; Gupta et al., 2002) and tetrasulfophthalocyanine (FePcS) (Meunier, 2002; Sorokin et al., 1995; Sorokin et al., 1996) were examined recently for there ability to activate H_2O_2. The hypothesized reason for using a tetradentate ligand is to allow for uncoordinated sites around the iron to be available for H_2O_2 activation. This has the effect of placing the H_2O_2 and Fe^{II} in close proximity, allowing for optimal kinetics. It would then seem counterintuitive to use a seven-coordinate ligand such as EDTA. Nevertheless, the activation of H_2O_2 by Fe^{II}EDTA to form reactive oxygen species is fully documented (Brausam and van Eldik, 2004; Feig et al., 1996; Seibig and van Eldik, 1997; Zang and van Eldik, 1990), and so is the production of hydroxyl radicals by Fe^{II}EDTA-O_2 systems (Yurkova et al., 1999).

The geometry of the $Fe^{II/III}$EDTA complex in solution is recognized to be an important factor in controlling reactivity at the iron center. $Fe^{II/III}$EDTA geometry is controlled by protonation of the EDTA chelate. Figures 5.7 and 5.8 show pH

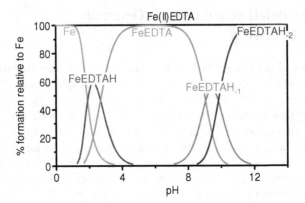

FIGURE 5.7. Distribution of $Fe^{II}EDTA$ species with respect to pH, showing predominately nonprotonated species from pH 4 to pH 8. (H_{-1} indicates OH group). Constants obtained from NIST; calculations performed by HySS software.

distribution diagrams of $Fe^{II}EDTA$ and $Fe^{III}EDTA$, respectively. Crystal structures of $[Fe^{II}EDTA(H_2O)]^{2-}$ at neutral pH show a seven-coordinate monocapped trigonal prismatic (MCP) structure with water in the seventh coordinate (Table 5.4 and Figure 5.9) (Mizuta et al., 1993). On acidification (pH 2–3), crystal analysis reveals that the protonated carboxylate arms of EDTA remain attached while the geometry shifts to an approximate pentagonal-bipyramidal (PB) (Mizuta et al., 1995). The shift in geometry from MCP to PB allows for a greater bite angle surrounding the coordinated water molecule presenting the oxygen/hydrogen peroxide molecule more space to enter the coordination sphere and substitute the coordinate water (Seibig and van Eldik 1997). Recent research on the auto-oxidation of $Fe^{II}EDTA$ to

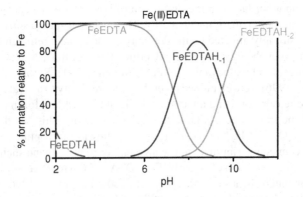

FIGURE 5.8. Distribution of $Fe^{II}EDTA$ species with respect to pH, showing predominately nonprotonated species from pH 2 to pH 6. (H_{-1} indicates OH group) Constants obtained from NIST; calculations performed by HySS software.

TABLE 5.4. The geometry and coordination of $Fe^{II}EDTA$ and $Fe^{III}EDTA$ complexes. Seven-coordinate complexes have water in the seventh coordinate and six-coordinate complexes in the sixth coordinate in the case of $[Fe^{III}(HEDTA)(H_2O)]$

	Coordinates	Structure	Reference
$[Fe^{II}(EDTA)(H_2O)]$	7	monocapped trigonal prism	Mizuta et al., 1993
$[Fe^{II}(HEDTA)(H_2O)]$	7	distorted pentagonal	Handshaw and Martell, 1998
$[Fe^{II}(H_2EDTA)(H_2O)]$		bipyramidal	
$[Fe^{III}(EDTA)(H_2O)]$	7	distorted pentagonal	Lopez-Alcala et al., 1984;
		bipyramidal	Solans and Altaba, 1984;
			Solans and Altaba, 1985
$[Fe^{III}(HEDTA)(H_2O)]$	6	distorted octahedral	Mizuta et al., 1990

$Fe^{III}EDTA$ by molecular oxygen (Seibig and van Eldik 1997) has shown increased kinetic rates at pH $\sim$3. The shift in geometry due to protonation (see Figure 5.3) and the consequent widening of the activation site are a possible explanation for the increased oxidation rate. pK_a values of 3.0, 2.8, 2.5, and 2.1 have been reported for the protonation of the EDTA ligand (Handshaw and Martell, 1998; Seibig and van Eldik 1997).

Geometry changes also occur in the ferric form of $Fe^{III}EDTA$ on acidification. The high-spin Fe^{II} ion has an ionic radius of 0.92 Å (Mizuta et al., 1995), which is large enough to take a coordination number of seven, while the Fe^{III} ion has a smaller radius of 0.785 Å (Mizuta et al., 1995), which is near the critical radius between coordination numbers six and seven. $[Fe^{III}(EDTA)(H_2O)]$ has been

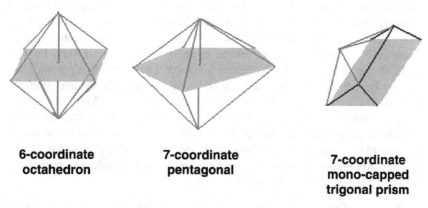

6-coordinate octahedron   **7-coordinate pentagonal**   **7-coordinate mono-capped trigonal prism**

FIGURE 5.9. General geometric shapes for six- and seven-coordinate $Fe^{II/III}EDTA$ complexes. Crystal structures given in the literature indicate in all cases a slightly distorted deviation from these shapes, with wider bite angles around the coordinated water and smaller angles associated with the chelate arms of EDTA. (Handshaw and Martell, 1998; Lopez-Alcala et al., 1984; Mizuta et al., 1990; Mizuta et al., 1993; Solans and Altaba, 1984; Solans and Altaba, 1985).

crystallized as a seven-coordinate distorted pentagonal bipyramid similar to the ferrous forms, although on acidification the geometry of $[Fe^{III}(HEDTA)H_2O]^-$ complex changes to six-coordinate distorted octahedral with one protontated acetate group freed from coordination and the water molecule occupying the sixth coordinate (Mizuta et al., 1995). The octahedral geometry is considerably distorted from a regular symmetric shape with angles associated with the coordinated water and Fe^{III} being much larger than 90° (103.1°–104.4°) (Mizuta et al., 1990). Again the distorted shape allows for a widening of the activation site, lowering the kinetic barrier for hydrogen peroxide and/or molecular oxygen activation.

5.6. Form of Reactive Oxygen Intermediate Species

Over the past two decades, much effort has been put forth to elucidate the reaction mechanism and intermediate reactive oxygen species present during the transformation of dioxygen by iron complexes in both abiotic and biotic systems. The reaction between $Fe^{II}EDTA$ and hydrogen peroxide has long been used as a model system to understand the mechanisms of more complicated iron systems (Neese and Solomon, 1998; Sharma et al., 2004; Walling, 1975). Several $Fe^{III}EDTA$-peroxide intermediate species have been proposed and characterized spectroscopically (Brausamand van Eldik, 2004; Feig et al., 1996; Seibig and van Eldik, 1997; Sharma et al., 2004; Zang and van Eldik, 1990) in aqueous media, although no mononuclear Fe^{III} peroxo species complex has been isolated as yet (Brausam et al., 2004). One well-known iron/peroxo complex, first reported in 1956 by Cheng and Lott (Walling et al., 1970), is the purple-colored ($\lambda = 520$ nm) $[Fe^{III}EDTA(\eta^2-O_2)]^{3-}$, which has been demonstrated to form by the addition of excess H_2O_2 and $Fe^{III}EDTA$ in basic aqueous solutions (Brausam and van Eldik, 2004; Neese and Solomon, 1998; Sharma et al., 2004). Experimental spectroscopic evidence indicates the η^2 side-on geometry to be the most likely binding mode, shown in Figure 5.10 (Neeseand Solomon, 1998). The iron/peroxo complex has been found to be unreactive toward organic media in alkaline solutions (pH >10.5) (Neese et al., 1998); however, it acts vigorously at lower pH leading to the decomposition of excess EDTA (Neese and Solomon, 1998; Walling, 1975) and other organics present. The addition of acid is thought to protonate the bound peroxide to form a high-spin Fe^{III}-hydroperoxide intermediate (Figure 5.6). There are three possible mechanisms for organic degradation in which the intermediate is involved (Neese and Solomon, 1998):

1. Cleavage of the O—O bond would yield high-valent iron species such as $Fe^{(IV)}=O$. Ferryl has been studied in depth in relation to heme-based systems and is believed to be the active intermediate responsible for organic oxidation in these systems. Therefore one could assume that the $Fe^{(IV)}=O$ may play a role in organic oxidation, although the stability to the $Fe^{(IV)}=O$ complex in a nonheme, mononuclear, and aqueous environment may be too small for this to be a viable pathway for the ZEA scheme.

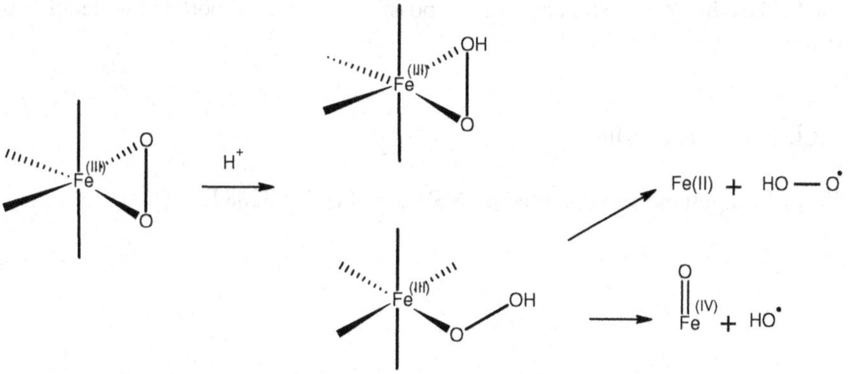

FIGURE 5.10. Proposed $[Fe^{III}EDTA(\eta^2-O_2)]^3$ dioxygen intermediate for $Fe^{III}EDTA$ systems.

2. The decay of the hydroperoxy complex could lead to the $Fe^{II}EDTA$ and the superoxide radical, which would initiate a radical chain reaction that could be propagated by typical Fenton chemistry, involving the OH• radical.

3. An alternative to both the mechanisms would be the iron/hydroperoxy intermediate itself acting as the reactive oxygen species; this would lead to more selectivity in the oxidation. This would be compatible with studies such as the Gif reaction that demonstrate a more selective oxidation than would be possible with OH• radicals.

While the ZEA reaction scheme does not involve the addition of H_2O_2 directly, there is recent evidence that it is produced in situ over the course of the reaction. The authors therefore believe that the most likely of the three mechanisms for the ZEA system is the second or the radical production mechanism because of the unselective nature of the ZEA oxidation scheme, although the other mechanistic possibilities cannot be ruled out. The data showing OH• formation during the autoxidation of $Fe^{II}EDTA$ to $Fe^{III}EDTA$ by O_2 (Seibig and van Eldik, 1997) and radical trap analysis on the ZEA system is evidence that radical production could play an important role in the ZEA system.

5.7. Conclusion

The uniqueness of the ZEA system in comparison to other advanced oxidation systems is its ability to activate oxygen at room temperature and pressure without the aid of expensive catalysts and solvents or costly energy requirements. The ZEA system is capable of degrading the phosphorus-sulfur groups present in malathion, as well as cleaving aromatic rings in chlorinated phenols to carbonates, and simple carboxylates in a one-step process eliminating the need for secondary oxidation processes. This characteristic, combined with inexpensive and stable reagents, establishes the ZEA system as a strong possibility as a field portable remediation system.

Acknowledgments

This investigation was supported by NSF award BES-0328827.

6
Quantification of the Oxidizing Capacity of Nanoparticulate Zero-Valent Iron and Assessment of Possible Environmental Applications

6.1. Introduction

In Chapter 4, it was shown that nanoscale zero-valent iron (nZVI) oxidizes the herbicide molinate when it is used in the presence of oxygen. Analysis of by-products observed during ZVI-mediated oxidative degradation of the herbicide is consistent with the action of a nonspecific oxidant such as the hydroxyl radical (OH$^{\bullet}$). The reactions that result in oxidant production are initiated when Fe^0 is oxidized by oxygen, which likely forms a reactive oxygen species either on the particle surface or in solution. The oxidation of Fe^0 by oxygen also results in the formation of a layer on the particle surface with properties similar to γ-Fe$_2$O$_3$ and Fe$_3$O$_4$, which eventually leads to passivation of the surface accompanied by a decrease in the rate of Fe^0 oxidation (Davenport et al., 2000). Passivation of the surface also appears to be responsible for the decreased rate of Fe^0-mediated oxidation of molinate that is observed over extended time. For a system involving granular Fe$^0_{(s)}$, the diminution of reactivity can be counteracted by the addition of a chelating agent (such as EDTA) that keeps the oxidized iron in solution (Noradoun et al., 2003). However, addition of chelating agents limits the utility of the technique for in situ treatment. The high surface area of nZVI may allow for more efficient generation of oxidants, but a decrease in reactivity associated with the buildup of iron oxides on the surface eventually slows the reaction. The corrosion of the Fe^0 particles is accompanied by observation of release of ferrous ions to solution and generation of hydrogen peroxide suggesting the possibility of a Fenton-like reaction at or near the particle surface. If the nZVI particles continue to produce oxidants after a surface coating forms, the continued slow release of oxidants may result in degradation of contaminants present in soil and contaminated aquifers. The high surface area of the nZVI particles could also provide a means for selective oxidation of surface-active compounds.

To assess the potential applicability of nZVI for the oxidative treatment of organic contaminants and to investigate the reaction mechanism and yields of the oxidizing species further, experiments were performed using the oxidation of benzoic acid (BA) to *p*-hydroxybenzoic acid (*p*-HBA) as a probe reaction for oxidant production. Analysis of oxidative nZVI reactions under well-defined conditions provides insight into the effect of environmental conditions (e.g., pH) on oxidation rates, the selectivity of the oxidant, and the overall efficiency of the process. Additionally, insights gained from the laboratory investigations under controlled conditions are used to assess the potential applications of this technology and to identify areas for further research. The results of the benzoic acid studies are presented in this chapter, as is an assessment of future practical applications of the oxidative use of nZVI.

6.2. Results

6.2.1. p-*Hydroxybenzoic Acid (*p-HBA*) Formation*

The *p*-isomer is the dominant oxidation product formed on oxidation of benzoic acid, and concentrations of this product generated as a function of time of reaction between benzoic acid and 0.9 mM nanosized Fe^0 are shown in Figure 6.1a. Concentrations of *p*-HBA formed for different initial concentrations of benzoic acid after a 2-h reaction time with 0.9 mM Fe^0 are shown in Figure 6.1b. The reaction goes to completion within the first hour with very little increase (at least at [benzoic acid]$_0$ = 10 mM) in concentration of the reaction product *p*-HBA over the 10 μM (or 0.1% of the initial benzoic acid concentration) achieved in the first hour. Increasing the initial concentration of benzoic acid results in an increase in the concentration of *p*-HBA produced, though the increase is less than proportional with the concentration of *p*-HBA plateauing at around 10 mM at the higher benzoic acid concentrations examined.

6.2.2. Cumulative Hydroxyl Radical Formation over Long Term

Results of longer term (1 day) studies of hydroxyl radical production are shown in Figure 6.2 and suggest that, although a little slow initially, nZVI-mediated oxidation will continue for some time, particularly in the mid-pH range. The production of *p*-HBA from BA exhibited biphasic kinetics, with rapid initial production of *p*-HBA followed by a slow increase over periods of at least 1 day (Figure 6.2). The concentration of *p*-HBA produced after 1 h was used in all subsequent experiments to compare the rates of oxidation under different conditions. Under the conditions used in these experiments (i.e., 10 mM BA; 0.9 mM nZVI), less than 5% of the BA initially present was transformed, and therefore it served as the main sink for oxidants with the result that oxidation of *p*-HBA was negligible. Thus, the concentration of *p*-HBA formed is related to the concentration of oxidizing species that could be used to transform a contaminant.

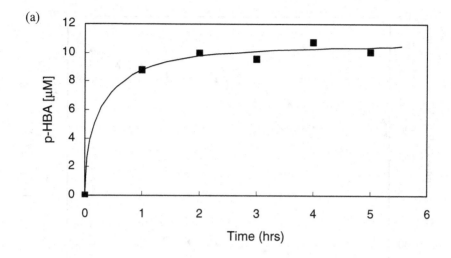

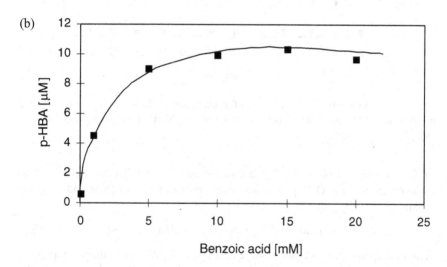

FIGURE 6.1. *p*-HBA formation over time in pH 3 suspension and 30 mM ionic strength containing 10 mM benzoic acid (a), and *p*-HBA formation after 2 h as a function of benzoic acid concentration (b).

BA is transformed into three isomers of hydroxybenzoic acid. Although it was not possible to quantify each of the isomers because of difficulty resolving the ortho and meta forms, the three isomers appeared to be present at similar concentrations. The three isomers of hydroxybenzoic acid account for $90 \pm 5\%$ of the products of OH• reactions with BA with the ratio of *o*-HBA, *m*-HBA, and *p*-HBA products reported to be in the proportion 1.7 : 2.3 : 1.2 (Klein et al., 1975). For the oxidation

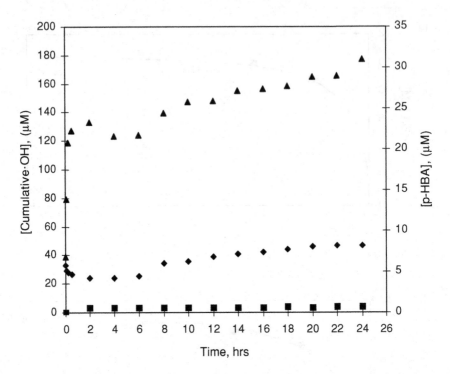

FIGURE 6.2. Cumulative hydroxyl radical formation and p-HBA concentration over time at pH 3 (▲), 5 (♦), and 8 (■) (Conditions: 0.9 mM Fe^0, 30 mM ionic strength, and 10 mM BA) (Joo et al., 2005).

of BA by solution phase OH•, the concentration of p-HBA has been used to estimate cumulative OH• production using Equation (6.1) (Zhou and Mopper, 1990):

$$\text{Cumulative OH• produced} = [p\text{-HBA}] \times 5.87 \qquad (6.1)$$

A similar approach is adopted with results from all nZVI experiments expressed in terms of both p-HBA concentration and cumulative OH• production, with the latter quantity estimated using Equation (6.1).

6.2.3. Effect of Fe(II) as Oxidant Scavenger

To assess competition for oxidants between BA and other oxidant scavengers, such as Fe(II), experiments were conducted at pH 3 over a range of BA concentrations. The yield of p-HBA increased as BA concentrations increased from 50 μM to 5 mM and then remained approximately constant up to 20 mM (Figure 6.3). Simultaneous measurements of ferrous iron indicated that total Fe(II) ranged from 190 to 200 μM (Figure 6.4) in almost all cases and was relatively constant over the duration of the experiment.

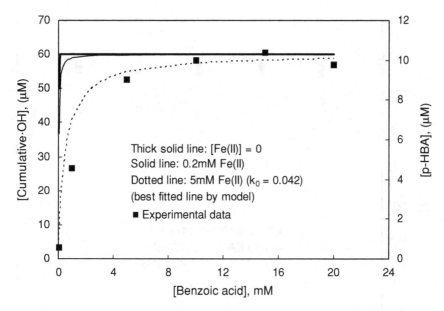

FIGURE 6.3. Oxidant formation as a function of BA concentration and predicted oxidant formation in the absence and presence of Fe(II) scavenging OH radicals (Conditions: 0.9 mM Fe^0, pH 3, 30 mM ionic strength, reaction time 2 h). Solid and dashed lines show calculated [p-HBA] assuming competitive hydroxyl radical scavenging by 0.2 and 5 mM Fe(II), respectively.

6.2.4. Effect of ZVI Concentrations on Oxidant Yield

Altering the relative concentrations of the target compound (benzoic acid) to those of ZVI can affect the extent of oxidation product formation (Figure 6.5). At low initial concentrations of benzoic acid (50 µM), the dependence is not strong and an increase in the concentration of ZVI does not substantially influence the quantity of p-HBA formed. A ninefold increase in ZVI concentration from 0.2 to 1.8 mM results in slightly less than a doubling in p-HBA concentration produced (Figure 6.5a). At even higher ZVI concentrations, the concentration of p-HBA produced is observed to decrease, possibly as a result of scavenging of oxidant by ZVI. Experiments with higher BA concentrations (10 mM) show a substantial increase in p-HBA production as nZVI increased (Figure 6.5b). The difference in p-HBA production for 0.2–5 mM ZVI is about 700%.

Assuming the same distribution of HBA isomers in the nZVI system, the results can be used to estimate the efficiency of the initial fast reaction (i.e., the yield of oxidants per mole of nZVI added). The reaction efficiency ranged from approximately 5 to 25%, with higher values at the lowest ZVI concentration (Figure 6.6). The effectiveness of the nZVI as an oxidant generator decreases on increasing nZVI loading suggesting the need to use a low dose of nZVI for maximum efficiency.

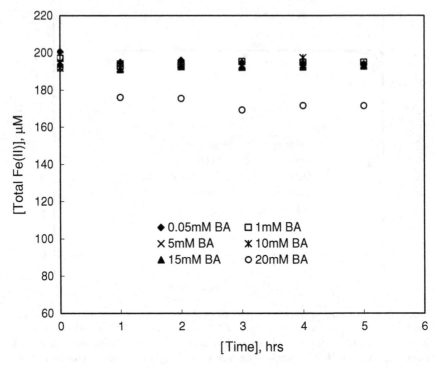

FIGURE 6.4. Total Fe(II) in suspension over time as a function of BA concentration at pH 3 and ionic strength of 0.03 M.

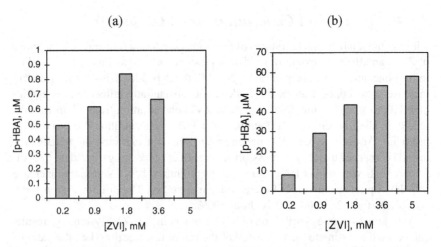

FIGURE 6.5. *p*-HBA formation as a function of ZVI concentration (Conditions: pH 3 solutions, 30 mM ionic strength, reaction time 1 h, and [benzoic acid]$_o$ = 50 μM in (a) and [benzoic acid]$_o$ = 10 mM in (b)).

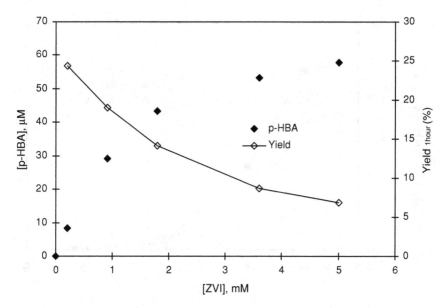

FIGURE 6.6. Effect of increase in mass of nZVI used on concentration of principal oxidized products (*p*-HBA) (Conditions: 10 mM BA, 30 mM ionic strength, pH 3, reaction time 1 h) (Joo et al., 2005).

6.2.5. Effect of pH

The oxidizing ability of nZVI (as measured by the extent of *p*-HBA production up to 1 hour) as a function of pH and relative to oxidizing ability at pH 3 is given in Figure 6.7 for two concentrations of BA, 50 μM and 10 mM. The pH is observed to significantly influence the oxidizing ability of nZVI, with a marked decrease in concentration of the oxidized product (*p*-HBA) with increasing pH. This effect of pH could be related to passivation of the ZVI surface at higher pH or to the presence of more effective hydroxyl radical deactivation pathways at the higher pH. Despite the decrease in oxidant production rate as pH increased, oxidation was observed at pH values of up to 8.

6.2.6. Selectivity of Oxidant

To elucidate the selectivity of the oxidant, and the importance of surface versus solution-phase reactions, competition experiments were conducted with aniline, *o*-hydroxybenzoic acid, phenol, and humic acid at pH 3 and 10 mM BA. In all cases, the yield of *p*-HBA decreased as the concentration of the competitor increased (Figures 6.8–6.11). The relative rate constant (i.e., the rate constant for each probe compound relative to the rate constant for BA) was estimated by least squares fit

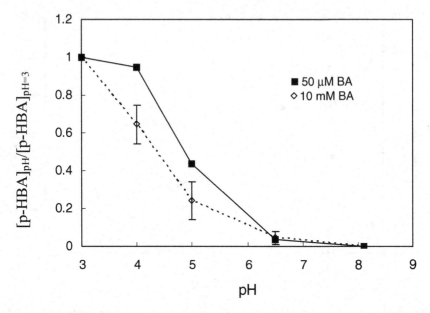

FIGURE 6.7. pH dependence of the formation of p-HBA (Conditions: pH 3, 0.9 mM ZVI, ionic strength 30 mM, and reaction time 1 h) (Joo et al., 2005).

of the data to Equation (6.2):

$$F = \frac{k_{BA}[BA]}{k_{BA}[BA] + k_C[C]} = \frac{1}{1 + \frac{k_C}{k_{BA}}\frac{[C]}{[BA]}} = \frac{1}{1 + k_{C/BA}\frac{[C]}{[BA]}} \quad (6.2)$$

where F is the fraction of the p-HBA produced in the presence of a concentration of a competitor (C) to that in the absence of C and $k_{C/BA}$ is the relative rate constant. Relative rate constants were also determined for the compounds in a similar manner but using Fenton's reagent as a source of OH$^\bullet$ (Figures 6.8b–6.11b). Experimentally determined relative rate constants for the four probe compounds are given in Table 6.1, as are the literature values for the absolute rate constants for reaction of the probe compounds with hydroxyl radicals.

The relative rate constants for aniline, a positively charged molecule at pH 3, are similar in both the nZVI and Fenton systems suggesting no effect of the nZVI surface on reaction rate. For phenol, the relative rate constant obtained with nZVI is twice as high as that obtained with hydroxyl radicals generated in the solution phase, while the relative rate constant for humic acid with nZVI-mediated oxidation is about one-third of that found with the Fenton reagent. Uncertainty surrounds the quality of the relative rate data for o-HBA since the BA/o-HBA solution turned purple on adding Fe(II), presumably because of complexation between BA or o-HBA and Fe(II). Consistent with the low relative rate constant obtained in the Fenton case, this complexation might be expected to retard the rate of reaction of Fe(II) with H_2O_2 and therefore prevent the generation of OH$^\bullet$.

(a)

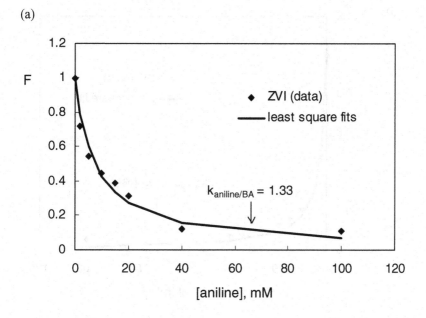

(b)

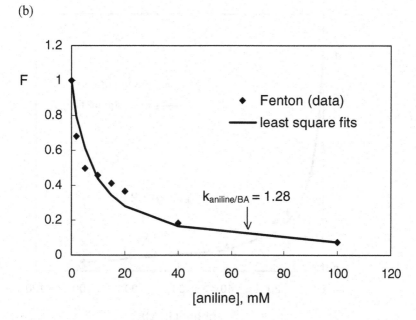

FIGURE 6.8. Fraction (F) of p-HBA produced in the presence of a certain concentration of aniline to that produced in the absence of aniline (i.e. in the presence of BA only) with (a) nZVI used as the oxidant source, and (b) Fenton reagent used as the oxidant source.

(a)

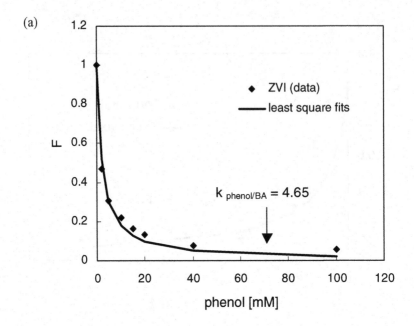

(b)

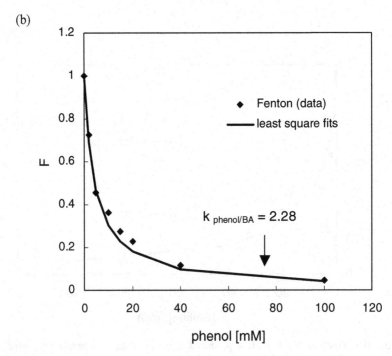

FIGURE 6.9. Fraction (F) of p-HBA produced in the presence of a certain concentration of phenol to that produced in the absence of phenol (i.e. in the presence of BA only) with (a) nZVI used as the oxidant source, and (b) Fenton reagent used as the oxidant source.

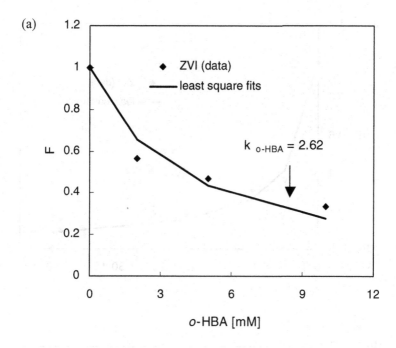

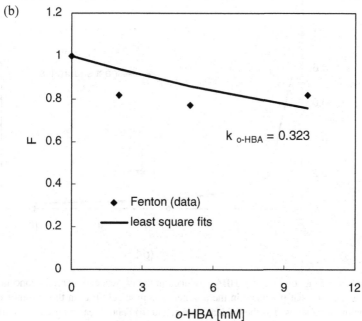

FIGURE 6.10. Fraction (F) of p-HBA produced in the presence of a certain concentration of o-HBA to that produced in the absence of o-HBA (i.e. in the presence of BA only) with (a) nZVI used as the oxidant source, and (b) Fenton reagent used as the oxidant source.

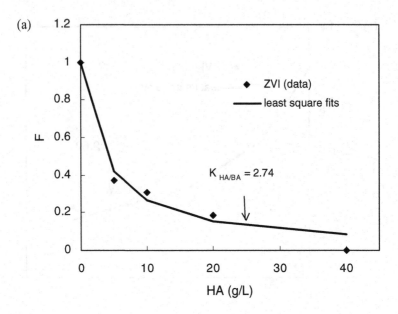

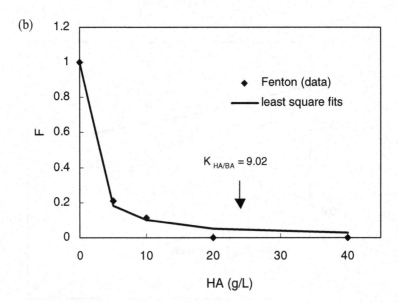

FIGURE 6.11. Fraction (F) of p-HBA produced in the presence of a certain concentration of humic acid to that produced in the absence of humic acid (i.e. in the presence of BA only) with (a) nZVI used as the oxidant source, and (b) Fenton reagent used as the oxidant source.

TABLE 6.1. Rate constants for reaction of probe molecules with hydroxyl radicals generated at pH 3 using both nZVI and Fenton reagents relative to that of benzoic acid ($k_{C/BA}$). The literature values for reaction of benzoic acid and probe compounds (C) with hydroxyl radicals are also shown.

Compound	pK_a	Relative rate constant ($k_{C/BA}$)		Reported rate constant for reaction with OH$^\bullet$ (k_C) (M^{-1}s^{-1})	Reference
		nZVI	Fenton		
BA	4.2	1.00	1.00	4.3×10^9 (k_{BA})	Buxton et al., 1988
Aniline	4.6	1.33	1.28	$8.6 \times 10^9 - 1.7 \times 10^{10}$	Rivas et al., 2001
Phenol	10.0	4.65	2.28	$6.6 \times 10^9 - 1.8 \times 10^{10}$	Rivas et al., 2001
o-HBA	3.0	2.62	0.32	$1.1 \times 10^{10} - 1.8 \times 10^{10}$	Rivas et al., 2001; Zečević et al., 1989
Humic acid	>3.6	2.74^a	9.02^b	–	–

^{a} Units of (L·M·s^2)/mg.
^{b} Units of L/mg/s.

6.2.7. Effect of ZVI Type on Oxidant Yields

The effect of the type of ZVI was evaluated at pH 3 with 10 mM BA. All four types of ZVI particles were capable of oxidizing BA, and p-HBA production increased as the total ZVI concentration increased (Figure 6.12). While the nZVI produced more p-HBA than the other forms of Fe^0, the rates of production for the other forms of Fe^0 were very significant, particularly in light of their comparative surface areas (Table 6.2).

6.2.8. Comparison Study on Standard Fenton Oxidation of Benzoic Acid

To compare with the nZVI system and to assess the oxidant yield by the Fenton process, the oxidation of benzoic acid using Fenton's reagents (e.g., 200 μM Fe^{2+} and 20 μM H$_2$O$_2$) was examined.

As can be seen in the Figure 6.13, a similar oxidation trend was seen in the range of 0.05–20 mM BA for both nZVI and Fenton's processes. Nevertheless, more oxidation at lower BA concentration (e.g. 0.05 mM) was seen in the nZVI system (Figure 6.13b) compared with the Fenton system, although this depends on the amount of iron used. As shown in Figure 6.14, the cumulative oxidant yield by Fenton was around 14 μM, which is lower than the expected oxidant yield of 20 μM based on the amount of Fe(II) added to the system. Since o- HBA was not detected during the reaction, but p- and m-HBA were, the Fe(II) might be complexing the o-HBA and inhibiting the Fenton reaction, consequently lowering its efficiency.

6.2.9. Effect of Pure O$_2$ on Oxidant Yield

The effect of replacing air with pure oxygen was investigated by sparging the ZVI suspension with oxygen gas, and the kinetics of oxidant yield were then assessed.

TABLE 6.2. Characteristics of ZVI particles used in this study

ZVI	Measured BET surface area $(m^2/g)^a$	Literature surface area (m^2/g)	Particle size $(u\mu m)$
Master Builders	0.71	1.3 ± 0.7 (Alowitz and Scherer, 2002)	750–1200
Aldrich	0.38	0.9 ± 1.1 (Alowitz and Scherer, 2002)	4.5–5.5
Kanto	0.60	–	100–150
nZVI^b	32	31.4 (Choe et al., 2001)	0.001–0.2

^{a} BET analyses performed on ZVI as received.
^{b} nZVI prepared by $NaBH_4$ reduction of $FeCl_3$ in this study.

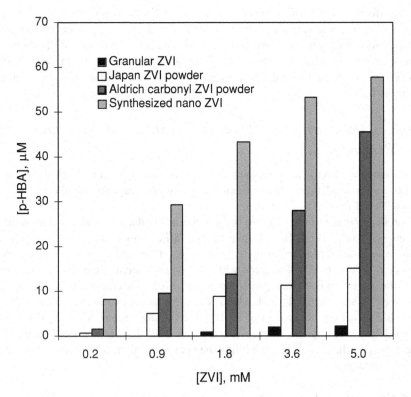

FIGURE 6.12. Concentration of p-HBA produced after a 1 h reaction of commercial granular Fe, ZVI powders, and synthesized nZVI with benzoic acid. (Conditions: pH 3, 0.9 mM Fe^0, ionic strength 30 mM, 10 mM BA, reaction time 1 h) (Joo et al., 2005).

(a)

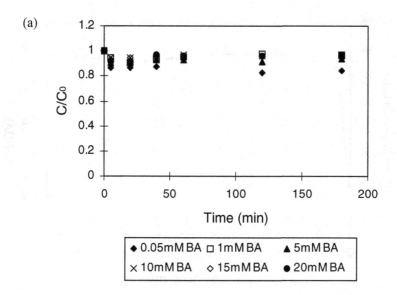

(b)

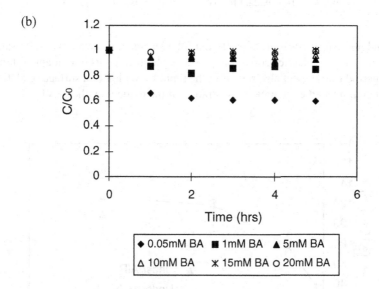

FIGURE 6.13. BA oxidation by Fenton's reagents (a) and ZVI (0.9 mM) (b) at pH 3.

As can be seen in Figure 6.15, the cumulative formation of hydroxyl radicals was greater under O_2 compared with an air-sparged system. Within 1 min, 88 μM compared with 40 μM of cumulative hydroxyl radicals are formed when sparging with O_2 compared with air. The rate of oxidant formation slowed considerably for the O_2 system after 1 min, and there was little difference in yield for the

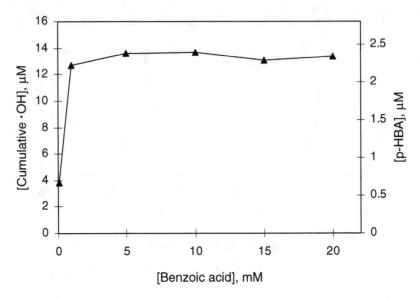

FIGURE 6.14. Oxidant formation in Fenton oxidation system.

O$_2$- and air-sparged systems after 10 min. The results indicate that O$_2$ supersaturation not only accelerates corrosion, producing high amounts of oxidant within a short period of time, but also results in the rapid formation of a surface oxidation layer that hinders the continuous generation of the oxidant by the nZVI.

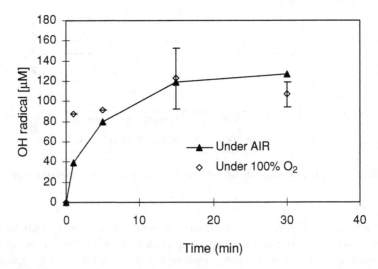

FIGURE 6.15. Kinetics of oxidant formation in nZVI system under O$_2$ and air sparging. Conditions: 10 mM BA, 0.9 mM ZVI, pH 3, 30 mM ionic strength.

6.2.10. Discussion

The first step in the oxidation of BA by nZVI involves the oxidation of Fe^0 by O_2. Chemists and material scientists have studied this process, which is referred to as *corrosion*, for many years. One interpretation of corrosion posits the initial oxidation of Fe^0 as a 2-electron process (Bozec et al., 2001; Zećević et al., 1989; Zećević et al., 1991):

$$Fe^0 + O_2 + 2H^+ \rightarrow Fe^{2+} + H_2O_2 \tag{6.3}$$

The H_2O_2 produced in Equation (6.3) could oxidize another Fe^0:

$$Fe^0 + H_2O_2 \rightarrow Fe(II) + 2OH^- \tag{6.4}$$

This process results in a 4-electron transfer and an overall stoichiometry of 2 mol of Fe^0 oxidized per mole of O_2. Alternatively, H_2O_2 could react with species such as Fe(II):

$$Fe(II) + H_2O_2 \rightarrow Fe^{3+} + OH^\bullet + OH^- \tag{6.5}$$

These reactions could occur on the Fe^0 surface or involve transfer of electrons through an iron oxide layer (Zećević et al., 1989), depending on the reaction rates and affinity of the species for surfaces. Although the 4-electron transfer process is usually the dominant mechanism of O_2-mediated corrosion, the 2-electron transfer process can result in the production of significant amounts of O_2^{2-} or H_2O_2, especially after an oxide coating has been formed on the surface. For example, in studies conducted on stainless steel rotating disc electrodes, 10–20% of the oxygen reduced during corrosion was converted into O_2^{2-} or H_2O_2 (Bozec et al., 2001).

On Fe^0 surfaces, the initial rate of the reaction is fast. However, a surface oxide layer consisting of γ-Fe_2O_3- and Fe_3O_4-like oxides forms as the surface is oxidized (Davenport et al., 2000). This passive layer protects the remaining Fe^0 from attack by oxygen by serving as a barrier between the Fe^0 and O_2. The initial pulse of oxidants produced in the nZVI system is consistent with oxidation of the surface Fe^0 serving as the source of oxidant. For example, assuming that each Fe^0 occupies an area equal to 1.6×10^{-19} m^2 (i.e., assuming an atomic radius of 1.4 Å and an Fe—Fe bond distance of 2.5 Å), a monolayer would correspond to approximately 1.8% of the total ZVI. Therefore, the maximum reaction efficiency for the initial first part of the reaction at pH 3 (Figure 6.6) would correspond to a corrosion layer that was about 14 molecules in thickness.

Reaction (6.4) provides a possible explanation for the observed oxidation of BA and other organic compounds that occurs when nZVI is exposed to O_2. The simplest model for this process would involve the production of $OH^\bullet$ in solution followed by a homogeneous reaction between BA and $OH^\bullet$. In this case, the rate at which BA is oxidized would depend on the relative concentrations of different $OH^\bullet$ sinks. Under the conditions in this system, the dominant sinks are expected to be

$$BA + OH^\bullet \rightarrow \text{intermediates}$$
$$k = 4.3 \times 10^9 M^{-1}s^{-1} \text{(Andreozzi and Marotta, 2004)} \tag{6.6}$$
$$Fe^{2+} + OH^\bullet \rightarrow Fe^{3+} + OH^-$$
$$k = 4.0 \times 10^8 M^{-1}s^{-1} \text{(Sychev et al., 1979)} \tag{6.7}$$

When an aromatic compound reacts with OH•, a cyclohexadienyl radical is formed that reacts with oxygen, eventually resulting in the production of hydroxylated products (Dorfman et al., 1962; Eberhard and Yoshida, 1973). At the total Fe(II) concentrations measured in the experiments described in Figure 6.3 (approximately 0.2 mM), benzoic acid should be a dominant sink for OH• if the reaction occurs in the bulk solution. As indicated by the calculated fit depicted in Figure 6.3, OH• is scavenged more effectively than predicted by only >5 mM Fe(II). Under the conditions used in these experiments, the pseudo-first-order rate constant for other unknown sinks would have to be equal to 2×10^6 s^{-1} (dashed line in Figure 6.3). If the sink is dissolved Fe(II), this would correspond to 5 mM Fe^{2+}.

An alternative explanation for the results depicted in Figure 6.3 is that a hydroxyl radical is generated on or adjacent to the surface. Under such conditions, Fe(II) adsorbed on the nZVI surface or concentrated in the area immediately adjacent to the surface could more effectively compete with BA for OH•. If this were the case, it would be expected that organic compounds with a higher tendency to adsorb to the surface would be oxidized more quickly than compounds with a low affinity for surfaces. At pH 3, the surface of the nZVI is positively charged (Figure 3.5). Therefore, positively charged solutes such as aniline should have a lower affinity for the surface than neutral solutes and solutes that can form surface complexes (e.g., BA or phenol). Despite the tendency of BA to exhibit a higher surface affinity than aniline, the relative rate constant for the aniline (relative to BA) is nearly identical for the nZVI system and the Fenton's reagent system, in which OH• is generated in solution (Figures 6.8–6.11, Table 6.1). The relative rates for the other solutes also do not exhibit strong selectivity for species that are expected to have a higher affinity for the surfaces.

The apparent low selectivity observed for nZVI contrasts with the high selectivity observed in TiO$_2$ photocatalysis, in which the oxidant is produced at the surface. For example, the trichloroacetate anion is preferentially degraded at low pH when the TiO$_2$ surface (pH$_{zpc}$ = 6.25) is positively charged, whereas the positively charged cation chloroethylammonium is degraded preferentially at high pH when the TiO$_2$ surface has a net negative charge (Kormann et al., 1991). The absence of a relationship between the charge of the compound and relative rates of oxidation in the nZVI system suggests that a direct interaction between the surface and the solute is not important. As indicated earlier, the dependence of the yield on BA concentration suggests that the oxidant is not produced in the bulk solution. Therefore, it is possible that the initial oxidation occurs close to the surface in a region in which the Fe(II) concentration is higher than the concentrations measured in the bulk solution. The detection of mixed Fe(II)/(III) oxides such as maghemite at the nZVI surface is consistent with the oxidation of Fe^{2+} to Fe(III) in these systems.

Irrespective of the identity of the oxidant or the location in which it is generated, these results indicate that oxidation of ZVI by oxygen results in the production of an oxidant that is capable of transforming organic contaminants. The initial pulse of oxidants might be useful in remediation of contaminated soils and groundwater and might serve as an alternative to other in situ treatment oxidation processes, such as the addition of H$_2$O$_2$. The slower release of oxidants that occurs after

30 min, which likely corresponds to the continued corrosion of the nZVI particle, might be useful for remediation of difficult-to-reach groundwater, provided that oxygen was not depleted from the groundwater. Additional research is needed to assess the kinetics of oxidant production from nZVI particles on the time scale of days to months.

The production of oxidants during Fe^0 oxidation is not a phenomenon that is limited to nZVI particles (Figure 6.12). Although researchers studying nZVI have noted the disappearance of O_2 as oxic water enters a Fe^0 permeable reactive barrier (O'Hannesin and Gillham, 1998; Yabusaki et al., 2001), the potential for transformation of contaminants in this zone has not been appreciated previously. The Fe-mediated production of oxidants on Fe^0 might provide new applications for ZVI reactive permeable barriers. For example, consider the leading edge of a groundwater plume that is contaminated with a highly mobile contaminant, such as methyl tertiary butyl ether (MTBE) or N-nitrosodimethylamine (NDMA). Assume that the plume initially is saturated with oxygen (i.e., 0.26 mM) and that it contains 1 µg/L (i.e., $\approx 10^{-8}$ M) of the contaminant, 2 mg/L of natural organic matter (i.e., $\approx 2 \times 10^{-6}$ M or less), and 1 mM of HCO_3^-. If all of the oxygen is consumed in the barrier by reaction with Fe^0 and 10% of the O_2 undergoes the 2-electron transfer pathway to form the oxidant, 26 µM of oxidant will be produced. Assuming that the oxidant exhibits the same reactivity as $OH^\bullet$, almost all of the contaminants would be removed in the reactive barrier via oxidative processes.

6.2.11. Conceptual Kinetic Modeling

6.2.11.1. ZVI Corrosion Process

Since the diffusion rate at steady state is proportional to oxygen concentration, it follows that the corrosion rate of iron is also proportional to oxygen concentration (Uhlig and Revie, 1985). Although an increase in oxygen concentration at first accelerates corrosion of iron, the corrosion rate drops again to a low value because of the formation of an oxide layer (Crow, 1984). The corrosion in the presence of oxygen is therefore initially very high and is largely dependent on the rate of diffusion of oxygen to the metal surface. The increased corrosion rate of iron as the pH decreases is not caused only by increased hydrogen evolution; in fact, greater accessibility of oxygen to the metal surface on dissolution of the surface oxide favors oxygen depolarization, which is often the more important reason (Uhlig and Revie, 1985).

Rapid corrosion may produce the initial pulse of peroxide through the $2e^-$ pathway in the presence of oxygen. However, such corrosion decreases as the reaction proceeds because of an oxide layer developing, which is from the $4e^-$ pathway. The continuous oxidant yield, degradation of molinate even at high pH, and increasing trend of oxidants at mid-pH over 24 h suggests that the $2e^-$ pathway, though transient, could continue through the corrosion reaction.

For processes that involve heterogeneous reactions, the apparent rate is usually dominated by either the rate of intrinsic chemical reactions on the surface or the rate of diffusion of the solutes to the surface (Uhlig and Revie, 1985). The average rate of reaction of H_2O_2 on the iron oxide surface is far slower than its diffusion rate to the surface through either the external film or internal pores (Lin and Gurol, 1998). Therefore, the intrinsic reactions on the oxide surface are expected to be the rate-limiting steps for this process. In general, the model for surface reactions consists of the following steps.

- Mass transport of the reactant to the Fe^0 surface from the bulk solution
- Adsorption of the reactant to the surface
- Chemical reaction at the surface
- Desorption of the products
- Mass transport of the products to the bulk solution.

It is proposed that the reactions occur at or near the Fe surface rather than in aqueous phase. However, probe compounds (aniline, phenol, o-HBA) showed similar relative rates of reaction, suggesting the key oxidation occurs in solution phase but near the iron surface.

Contaminant molecules in the aqueous phase have access to the iron surface where they can either adsorb to nonreactive sites or reach reactive sites where the reactions can take place. It is postulated that iron sorbed to the nonreactive sites cannot leave the surface and there is continuous regeneration of reactive sites as compounds degrade. Kitajima et al. (1978) proposed a possible mechanism for the heterogeneous goethite-catalyzed reaction (in the presence of peroxide) as described by

$$H_2O_2 + S \rightarrow OH^\bullet + OH^- + S^+$$

where S is the mineral surface and S^+ is an oxidized region of the surface. Consumption of peroxide increased as a function of its initial concentration, and the treatment is most economical at low H_2O_2 concentrations because of minimal quenching and more efficient stoichiometry. Based on homogeneous rate constants, the oxidation rate of ferrous sites by H_2O_2 (reaction (6.8)) is expected to be ~700 times faster than the oxidation rate of ferrous sites by oxygen (reaction (6.9)).

$$\equiv Fe^{II} + H_2O_2 \rightarrow \; \equiv Fe^{III}{-}OH + {}^\bullet OH + H_2O \qquad (6.8)$$

$$\equiv Fe^{II} + O_2 \rightarrow \; \equiv Fe^{III}{-}OH + HO_2^\bullet \qquad (6.9)$$

6.2.11.2. Key Reactions Involved in the ZVI Oxidation Process

The reduction of oxygen plays a key role in the corrosion of iron in aerated environments, and H_2O_2 has been found to be a reduction product (Bozec et al., 2001). The reduction of oxygen and the peroxide formation as an intermediate occur on the oxidized surface (Zečević et al., 1991). On the oxide-free surface the $4e^-$ reduction was found, although the kinetics of the O_2 reduction reaction is slower on more oxidized surfaces, with little hydrogen peroxide as an intermediate (less than 0.5% of the total reduction current; Jovancicevic and Bockris, 1986; Zečević et al.,

1989) and formation of superoxide radical, while on passive iron, oxygen reduction proceeds through a $2e^-$ pathway with the formation of hydrogen peroxide as a reaction product (Jovancicevic and Bockris, 1986; Zečević et al., 1989). The $4e^-$ pathway appears to proceed via the formation of adsorbed peroxide or superoxide where these adsorbates do not lead to a solution phase species (Jovancicevic and Bockris, 1986).

Oxygen adsorption occurs on iron with the rate of adsorption of oxygen expected to be low. Ferro et al. (2004) showed that the O_2 molecule only slightly physisorbs to a graphite surface and induces very little charge transfers. The rate constant for Equation (6.10) is unknown.

$$Fe^0 + O_2 \rightarrow\ >O_2 \quad k = \text{unknown} \tag{6.10}$$

The rate of oxidation of ZVI surface groups to >Fe(II) is also unknown.

$$Fe^0 +\ >O_2 \rightarrow\ >Fe^{2+} + 2>e^- \quad k = \text{unknown} \tag{6.11}$$

The reduction of $>O_2$ occurs rapidly as indicated by Bielski et al. (1985).

$$>O_2 +\ >e^- \rightarrow\ >O_2^{\bullet-} \quad k = 2 \times 10^{10}\ [M^{-1}s^{-1}] \tag{6.12}$$

Desorption of superoxide into solution occurs with a rate constant of 10^{-4} $[M^{-1}s^{-1}]$ estimated from modeling by Feitz and Waite (2003), although it was not measured experimentally.

$$>O_2^{\bullet-} \rightarrow O_2^{\bullet-} \quad k = 10^{-4}\ [M^{-1}s^{-1}] \tag{6.13}$$

The superoxide in solution results in the formation of H_2O_2 via the following reaction with rate constants at different pHs (Bielski et al., 1985):

$$HO_2^{\bullet} + HO_2^{\bullet} \rightarrow H_2O_2 + O_2 \quad k = 9.7 \times 10^7\ [M^{-1}s^{-1}]\ \text{at} \leq \text{pH} \leq 4.8 \tag{6.14a}$$

$$HO_2^{\bullet} + O_2^{\bullet-} + H_2O \rightarrow H_2O_2 + O_2 + OH\ k = 6.1 \times 10^4\ [M^{-1}s^{-1}]\ \text{at pH 8} \tag{6.14b}$$

Reaction (6.14a) is likely to occur relatively rapidly at low pH but more slowly at higher pH, resulting in the formation of hydrogen peroxide at or near the ZVI surface.

Ferrous iron released from ZVI might also react with the superoxide formed on the ZVI surface (Chen and Pignatello, 1997),

$$Fe^{2+} +\ >O_2^{\bullet-}(+2H^+) \rightarrow Fe^{3+} +\ >H_2O_2 \quad k = 1 \times 10^7\ [M^{-1}s^{-1}] \tag{6.15}$$

It might also be scavenged by hydroxyl radicals (Buxton et al., 1988) and superoxide in solution (Keene, 1964).

$$Fe^{2+} + OH^{\bullet} \rightarrow Fe^{3+} + OH^- \quad k = 4.3 \times 10^8\ (M^{-1}s^{-1}) \tag{6.16}$$

$$Fe^{2+} + O_2^{\bullet-} \rightarrow Fe^{3+} + HO_2^- \quad k = 7.3 \times 10^5\ [M^{-1}s^{-1}] \tag{6.17}$$

Ferrous iron is regenerated by the reaction between Fe(III) and $O_2^{\bullet-}$ (Nadezhdin et al., 1976):

$$Fe^{3+} + O_2^{\bullet-} \rightarrow Fe^{2+} + O_2 \quad k = 1.9 \times 10^9 \; [M^{-1}s^{-1}] \tag{6.18}$$

Ferrous oxidation and deactivation of superoxide are likely to be important factors affecting organic contaminants removal rate. It is likely that ZVI will be regenerated by the release of >Fe(II) to solution (as Fe(II)) though this is expected to occur relatively slowly with the rate constant estimated to be between 10^{-1} and 10^{-5} (s^{-1}) based on modeling studies by Feitz and Waite (2003) for desorption from TiO_2 surface.

$$> Fe^{2+} \rightarrow Fe^{2+} + \text{regenerated ZVI} \quad k = \text{unknown} \tag{6.19}$$

Since surface area concentrations of ZVI change with time, the surface area concentration is estimated assuming that the ZVI particles are spherical and the size is uniform as indicated by Chen et al. (2001).

$$\rho_{a(t)} = \rho_{a(0)}[M_t/M_0]^{2/3} \tag{6.20}$$

where $\rho_{a(0)}$ is the initial surface concentration, M_0 is the initial ZVI mass, and $M_{(t)}$ is the ZVI mass at time t. $M_{(t)}$ is calculated by subtracting the mass of total dissolved iron (i.e. ferrous iron concentration times the solution volume) from the initial iron loading. Chemical reactions occur on reactive sites, which are those where the breaking of bonds in the reactant molecules takes place, while on nonreactive sites only sorption interactions occur and the solute molecule remains intact (Burris et al., 1995).

Because of several unknown rate constants, the exact reactions are not drawn; however, it was possible to estimate hydroxyl radical production in the presence of oxygen using the data presented in Figure 6.3. The following reactions were used to assess the effect of Fe(II) presence on the cumulative generation of OH radicals as the function of the model compound (BA). The reactants and their concentrations used in the model using the ACHUCHEM kinetics program are as follows: ZVI, 0.9 mM; O_2, saturated 0.26 mM; BA, 5 μM–20 mM; Fe^{2+}, verified.

Reaction (6.21) is a simplification but represents the rate of generation of hydroxyl radicals at the ZVI surface from the interaction of ZVI surface sites and O_2.

$$Fe^0 + O_2 \rightarrow OH^{\bullet}$$
$$k = \text{unknown, but 0.042 is the optimized rate constant} \tag{6.21}$$
$$BA + OH^{\bullet} \rightarrow p\text{-HBA}$$
$$k = 4.3 \times 10^9 \text{(Andreozzi and Marotta, 2004)} \tag{6.22}$$
$$Fe^{2+} + OH^{\bullet} \rightarrow Fe^{3+} + OH^- \quad k = 4.3 \times 10^8 \tag{6.23}$$

The fitted rate constants for reaction (6.21) range from 0.04 to 0.042 $[M^{-1}s^{-1}]$, with the Fe(II) concentration of 4–5 mM. The optimized rate constant of reaction (6.21) is 0.042 $[M^{-1}s^{-1}]$ and the best fit scavenger concentration of Fe(II) with the rate constant is found to be 5 mM. As an example with benzoic acid, the peroxide formed through $2e^-$ oxygen reduction pathway reacts with ferrous iron via the Fenton reaction, producing hydroxy radical that can indiscriminately react with most organics in solution phase (Wells and Salam, 1968a).

$$Fe^{2+} + H_2O_2 \rightarrow Fe^{3+} + OH^{\bullet} + OH^- \quad k = 586 \; (pH \geq 4) \tag{6.24}$$

On the iron surface, adsorbed benzoic acid competes for hydroxyl radical formed on surface and/or interface near surface layer (Equation (6.23)). Although the Fenton reaction occurs in solution, it is likely that most primary reactions arise near the iron surface.

The hydroxyl radicals formed react with H_2O_2 (Sychev and Isak, 1995); for the decomposition of H_2O_2 (Bielski and Cabelli, 1995), the speciation rate constants (Bielski and Cabelli, 1995) are as follows.

$$OH^{\bullet} + H_2O_2 \rightarrow HO_2^{\bullet} + H_2O \qquad k = 3 \times 10^7 [M^{-1}s^{-1}] \tag{6.25}$$
$$H_2O_2 \rightarrow HO_2^- + H^+ \qquad k = 5 \times 10^{-2}(s^{-1}) \tag{6.26}$$
$$O_2^{\bullet-} + H^+ \rightarrow HO_2^{\bullet} \qquad k = 5 \times 10^{10}[M^{-1}s^{-1}] \tag{6.27}$$
$$HO_2^{\bullet} \rightarrow O_2^{\bullet-} + H^+ \qquad k = 5 \times 10^5(s^{-1}) \tag{6.28}$$
$$HO_2^- + H^+ \rightarrow H_2O_2 \qquad k = 5 \times 10^{10}[M^{-1}s^{-1}] \tag{6.29}$$

In the system containing bicarbonate, the rate constant is slower than that from most organic reactions with hydroxyl radicals (Buxton et al., 1988; Bielski et al., 1985):

$$HCO_3^- + OH^{\bullet} \rightarrow CO_3^{\bullet-} + H_2O \quad k = 8.5 \times 10^6 \; [M^{-1}s^{-1}] \tag{6.30}$$
$$CO_3^{\bullet-} + O_2^{\bullet-} \rightarrow CO_3^{2-} + O_2 \quad k = 6.5 \times 10^8 \; [M^{-1}s^{-1}] \tag{6.31}$$

The rate of catalytic decomposition of H_2O_2 on passive iron is very low, less than 8×10^{-11} mol cm^{-2} s^{-1} (Calvo and Schiffrin, 1984). The oxygen reduction is controlled by the properties of the surface and significantly influenced by the surface oxides (Bozec et al., 2001). In the four-electron pathway, O_2 reduction is limited by mass transport in the solution. It may be due to a limited access of oxygen to the metal surface, while on passivated surfaces, O_2 reduction occurs simultaneously with the reduction of ferric oxide. O_2 reduction rate is lower on oxide-covered surfaces than on bare metal electrode (Bozec et al., 2001).

A summary of the reactions that may be involved in the ZVI-mediated oxidation process, with rate constants from the literature values, is given in Table 6.3.

6.3. Conclusion

Addition of nanoparticulate zero-valent iron (nZVI) to oxygen-containing water results in oxidation of organic compounds. Additional studies on the oxidation of benzoic acid suggest that the oxidation involves the production of hydroxyl radicals through the reaction of ferrous iron and hydrogen peroxide. When nZVI was added to BA-containing water, an initial pulse of p-HBA was detected during the first

TABLE 6.3. Model reactions and rate constants (example of compound: benzoic acid)

Reactions	Constants ($M^{-1}s^{-1}$)	References
1. $Fe^0 + O_2 \rightarrow {>}O_2$	unknown	—
2. $Fe^0 + {>}O_2 \rightarrow {>}Fe^{2+} + 2 {>}e^-$	unknown	—
3. ${>}O_2 + {>}e^- \rightarrow {>}O_2^{\bullet -}$	2×10^{10}	Bielski et al. (1985)
4. ${>}O_2^{\bullet -} \rightarrow O_2^{\bullet -}$	1×10^{-4}	Feitz and Waite (2003)
5a. $HO_2^{\bullet} + HO_2^{\bullet} \rightarrow H_2O_2 + O_2$	$9.7 \times 10^7 (\leq pH\ 4.8)$	Bielski et al. (1985)
5b. $HO_2^{\bullet} + O_2^{\bullet -} + H_2O \rightarrow H_2O_2 + O_2 + OH^-$	6.1×10^4 (pH 8)	Bielski et al. (1985)
6. $Fe^{2+} + O_2^{\bullet -} \rightarrow Fe^{3+} + H_2O_2$	1×10^7	Chen and Pignatello (1997)
7. $Fe^{2+} + OH^{\bullet} \rightarrow Fe^{3+} + OH^-$	4.3×10^8	Buxton et al. (1988)
8. $Fe^{2+} + O_2^{\bullet -} \rightarrow Fe^{3+} + HO_2^-$	7.3×10^5	Keene (1964)
9. $Fe^{3+} + O_2^{\bullet -} \rightarrow Fe^{2+} + O_2$	1.9×10^9	Nadezhdin et al. (1976)
10. ${>}Fe^{2+} \rightarrow Fe^{2+} +$ regenerated ZVI	unknown	—
11. Benzoic acid + $OH^{\bullet} \rightarrow$ intermediates	4.3×10^9	Andreozzi and Marotta (2004)
12. $Fe^{2+} + H_2O_2 \rightarrow Fe^{3+} + OH^{\bullet} + OH^-$	586 (pH $\geq$ 4)	Wells and Salam (1968a)
13. $OH^{\bullet} + H_2O_2 \rightarrow HO_2^{\bullet} + H_2O$	3×10^7	Sychev and Isak (1995)
14. $H_2O_2 \rightarrow HO_2^- + H^+$	$5 \times 10^{-2} (s^{-1})$	Bielski and Cabelli (1995)
15. $O_2^{\bullet -} + H^+ \rightarrow HO_2^{\bullet}$	5×10^{10}	Bielski and Cabelli (1995)
16. $HO_2 \rightarrow O_2^{\bullet -} + H^+$	$5 \times 10^5 (s^{-1})$	Bielski and Cabelli (1995)
17. $HO_2^- + H^+ \rightarrow H_2O_2$	5×10^{10}	Bielski and Cabelli (1995)
18. $HCO_3^- + OH^{\bullet} \rightarrow CO_3^{\bullet -} + H_2O$	8.5×10^6	Buxton et al. (1988)
19. $CO_3^{\bullet -} + O_2^{\bullet -} \rightarrow CO_3^{2-} + O_2$	6.5×10^8	Bielski et al. (1985)

Note: Some rate constants were drawn from homogeneous studies. Therefore, caution must be exercised because the heterogeneous rate constants may be different.

30 min, followed by the slow generation of additional p-HBA over periods of at least 24 h. The yield of p-HBA increased with increasing BA concentration, presumably because of the increasing ability of BA to compete with alternate oxidant sinks, such as ferrous iron. At pH 3, maximum hydroxyl radical yields during the initial phase of the reaction were as high as 25% based on p-HBA measurements.

The initial rate of nZVI-mediated oxidation of BA exhibited a marked reduction at pH values above 3. Despite the decrease in oxidant production rate, p-HBA was observed during the initial reaction phase at pH values up to 8. Competition experiments with probe compounds expected to exhibit different affinities for the nZVI surface (phenol, aniline, o-hydroxybenzoic acid and synthetic humic acids) indicated relative rates of reaction that were similar to those observed in competition experiments in which hydroxyl radicals were generated in solution. A number of key findings, which have implications to the potential applications of this technology, are summarized below.

- Nanoscale zero-valent iron (nZVI) particles are capable of producing highly reactive and unselective hydroxyl radicals in the presence of oxygen. This newly discovered advanced oxidation technology therefore has the potential to oxidize almost all organic contaminants.
- As is the case for hydroxyl radicals, the presence of compounds other than the contaminant being targeted may reduce the efficiency of the process as a result of competition for the oxidant produced.

- While the process appears to exhibit high, short-term activity (especially at low pH), the process does appear to continue to be an effective oxidant generator over longer time periods and over a range of pH.
- The findings on competition study suggest that although the key reagents leading to generation of the powerful oxidants are sourced from the nZVI particles, the key oxidation step occurs in solution but in close proximity to the particle surface. The results of these studies will suggest possible applications of nanosized Fe^0 particles in oxidatively degrading chemicals of concern.
- The possibility that nanoscale iron particles can induce oxidative degradation of contaminants in the presence of oxygen markedly broadens the applicability of ZVI as a potentially useful reagent for the degradation of contaminants in waters and wastewaters.
- The nonspecificity of the nZVI oxidative process would appear to limit its application to scenarios involving high contaminant concentrations in systems where oxygen supply can be maintained.
- Consideration should be given to the possibility that ZVI-mediated oxidative processes are occurring at the surface of Fe^0 reactive barriers.
- Further investigations are required to both clarify the reaction mechanism and optimize operating conditions.

7
Conclusions and Future Research Needs

7.1. Column Studies

The discovery that colloidal particles of zero-valent iron (ZVI) in the presence of oxygen induce oxidative degradation of contaminants holds great promise as a technology for the degradation of contaminants in waters and wastewaters. The effectiveness of ZVI for the degradation of contaminants in drinking waters and recycled wastewaters was investigated in continuous column studies (Feitz et al., 2005). Three configurations (e.g., mixed ZVI/sand system; the same quantity of sand, gravel, and ZVI arranged so that the ZVI sits as a thin layer on top of the sand; ZVI supported on approximately 3 cm of sand that in turn was supported on loosely arranged gravel) were tested each with the same amount of ZVI (i.e. 0.2 g). Of these, the mixed ZVI/sand system was completely ineffective for the degradation of 100 ppb molinate under continuous conditions. In contrast, when the same quantity of sand, gravel, and ZVI was arranged so that the ZVI sat as a thin layer on top of the sand, continuous degradation was observed when compared with the control, but the efficiency in degradation decreased over the 3-h trial. The final configuration tested had ZVI supported on approximately 3 cm of sand that in turn was supported on loosely arranged gravel. This was the optimum configuration tested resulting in greater than 90% removal of molinate throughout the 3-h trial. The reasons for the substantial differences in degradation performance for the different configurations may be linked to oxygen availability. For example, in the mixed ZVI/sand system the embedded ZVI particles remained a black color throughout the trial, but in the ZVI-on-sand configuration the top of the ZVI layer turned a bright orange color, characteristic of Fe^0 oxidation. This indicated that oxygen was reacting at the interface between the molinate solution above the bed and the ZVI layer. In the optimum configuration, ZVI on sand/gravel, air trapped between the gravel may have ensured sufficient dissolved oxygen during the trial.

7.2. Further Applications of the ZVI-Mediated Oxidative Process

There are wide-ranging possible applications of the new discovery—it could be used in agriculture, industry, and water treatment for domestic use. For example, it can be applied to the filter cell by positioning nanoscale ZVI (nZVI) across soil, as can be seen in the Figure 7.1.

The action of filtration through the soil and nZVI purifies contaminated water. It could also complement biodegradation of contaminated water—when naturally occurring bacteria and the action of filtration through the soil purify contaminated water. Removing arsenic from water is another potentially important application of the technology. The treatment of arsenic in drinking water is very difficult to do using the oxidation reaction, so there is the potential that the nanoparticulate iron would serve as the oxidant of the arsenic. For example, the oxidation of As (III) to As (V) is involved in the nZVI removal mechanism as follows (Kanel et al., 2005):

$$Fe^0 + O_2 + 2H^+ \rightarrow Fe^{2+} + H_2O_2 \tag{7.1}$$

$$Fe^{2+} + H_2O_2 \rightarrow Fe^{III}OH^{2+} + OH\bullet \tag{7.2}$$

$$2OH\bullet + H_3AsO_3 \rightarrow H_2AsO_4^- + H_2O + H^+ \tag{7.3}$$

Nanoparticulate iron can also be used as part of methods for decontaminating surfaces of a solid substrate by applying to the surface particles of a zero-valent metal capable of reacting with oxygen and water to form hydroxyl radicals, and exposing the particles to oxygen and water, which can be carried out in the presence of air (Feitz et al., 2005).

Nanoscale iron particles may be also useful for a wide range of environmental applications. There are more flexible emplacement options for nanoscale iron

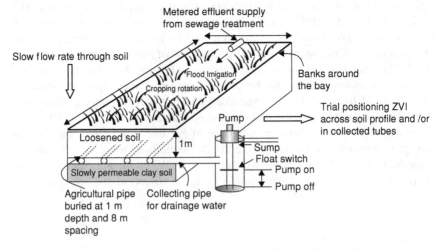

FIGURE 7.1. Potential application of nanoscale ZVI to soil using filter cell.

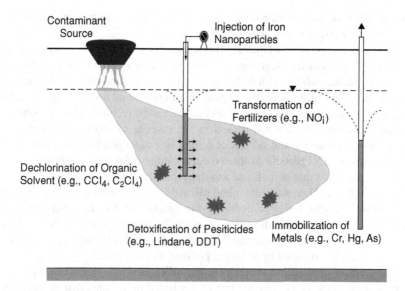

FIGURE 7.2. Schematic of nanoscale iron particles for in situ remediation (Zhang, 2003).

compared with granular iron and its use in permeable reactive barriers. For surface and groundwater remediation, the particles could remain suspended under very gentle agitation in an aqueous solution or could be injected into contaminated soils, sediments, and aquifers for in situ remediation of contaminants, as can be seen in Figure 7.2 as one of examples. Alternatively they could be anchored onto a solid matrix to enhance the treatment of water, wastewater, or gaseous process streams.

While the standard Fenton process has been reported to be relatively ineffective at higher pH, a high degree of degradation of contaminants was observed at alkaline pH in the radical-mediated process that is initiated by nZVI in the presence of dissolved oxygen (Joo et al., 2004). Degradation is possible at high pH and there is ongoing, although slower, decontamination after an initial rapid pulse of degradation activity (Joo et al., 2005). The process is effective as long as Fe^0 is available. This ongoing activity could enable the degradation of contaminants present in oxygen-rich groundwaters.

7.3. Summary of Results

Contamination of water and soil with agrochemicals is a global environmental problem. Pesticide residues have been detected in various natural waters in many countries and the presence of agrochemicals in drinking water supplies is of particular concern. It is an emerging problem in developing countries, and there is a genuine need for efficient and cost-effective remedial technologies. Thus, the investigation of remediation technology for polluted waters containing trace amounts of herbicides is of environmental interest.

The effectiveness of nZVI in degrading selected organic contaminants was found to vary widely. Extensive degradation was observed for aldrin, molinate, chlorpyrifos, and atrazine, while dieldrin, heptachlor, chlordane, and endosulfan were not degraded significantly under the experimental conditions used. Of the contaminants investigated, molinate was found to degrade effectively in oxic solutions containing nZVI, but removal ceased in the absence of oxygen. Molinate was subsequently chosen as the subject of further investigations into the kinetics and mechanism of nZVI particle facilitated degradation under oxic conditions.

The degradation rate of molinate was dependent on a number of physical conditions including ZVI concentration and pH. Additional studies using a ZVI/H_2O_2 process were highly effective, reducing the quantity of ZVI necessary for rapid removal of molinate. The degradation by-products were found to be the same for ZVI-alone, ZVI/H_2O_2, and the Fenton process ($Fe(II)/H_2O_2$), suggesting that molinate may be oxidized by the same pathway in each case.

The primary degradation products of the nZVI-mediated degradation process, keto-molinate isomers, have been previously found to be generated by hydroxyl-radical attack on molinate. That hydroxyl radicals or hydroxyl radical-like entities were involved in the degradation process is supported by findings that the hydroxyl radical scavenger, 1-butanol, was capable of completely limiting the oxidation of molinate. In addition, results of the reaction of molinate and nZVI in the presence of catalase (which removes H_2O_2) and in the presence of superoxide dismutase (which enhances the rate of formation of H_2O_2) lend further support to an oxidative mechanism involving hydrogen peroxide as an intermediate in the degradation pathway. Both hydrogen peroxide and Fe(II) were identified as reaction products.

The primary oxidant is therefore consistent with hydroxyl radicals. The radicals are likely to be formed as a result of reduction of oxygen to superoxide at the particle surface, disproportionation of the superoxide to hydrogen peroxide, and a Fenton reaction between the hydrogen peroxide and Fe(II) released during the corrosion process.

To identify the reaction mechanism and the factors that affect nZVI reactivity, a series of experiments were conducted using the formation of hydroxybenzoic acid (HBA) from benzoic acid (BA) as a probe for measuring the oxidizing capacity of nZVI. The p-isomer is the dominant product formed on oxidation of BA. The reaction went to completion within the first hour with little increase in reaction product concentration thereafter.

Increasing the initial concentration of BA results in an increase in the concentration of p-HBA produced. In contrast, the radical production efficiency of the nZVI as an oxidant generator decreases with increased nZVI loading, which suggests the need to use a low dose of nZVI for maximum efficiency. It was observed that almost 25% of the ZVI at low loading is converted to oxidant using the hydroxyl radical conversion factor noted by Zhou and Mopper (1990). The rate and overall yield of the reaction decreased with increasing pH, presumably because of either changes in the corrosion process or increasingly effective passivation effects.

Results of longer term (1 day) studies of hydroxyl radical production indicate that nZVI-mediated oxidation, although slow, will continue for some time, particularly in the low- to mid-pH range. The findings using competing probe molecules (phenol, aniline, and o-HBA) to ascertain the importance of surface versus solution phase reactions suggest that although the key reagents leading to generation of the powerful oxidants are sourced from the nZVI particles, the key oxidation step occurs in solution but presumably in close proximity to the particle surface given the highly reactive nature of hydroxyl radicals.

7.4. Overview of nZVI Research and Further Research Needs

The use of nZVI for remediation provides fundamental research opportunities and technological applications in environmental engineering and science. ZVI has proven to be useful for reductively transforming or degrading numerous types of organic and inorganic environmental contaminants. Few studies, however, have investigated the oxidation potential of ZVI. Contaminants investigated so far include molinate, atrazine, aldrin, chlorpyrifos from this study, methyl $tert$-butyl ether (MTBE) (by coupled ZVI/H_2O_2) from the finding by Bergendahl and Thies (2004), and possibly many nitrogen-containing NDMA precursors, including dimethylamine and trimethylamine, which might be suitable for oxidative degradation.

ZVI could be very useful for the large-scale remediation of organic contaminants because of the relative simplicity of the technology and the potential ready availability of key reactants. Since nanoscale iron particles have high surface reactivity with high surface areas, nZVI provides a new option in environmental remediation technologies and may provide cost-effective treatment to some of the most challenging environmental problems.

While such nZVI technology appears to be interesting, some issues need to be addressed and require further research. These issues are listed below.

1. It was suggested that a pH increase will favor the formation of iron hydroxide, $Fe(OH)_3$, which may eventually form a surface layer that inhibits iron dissolution Formation of other mineral phases can also occur depending on the ions available in solution. The formation of surface films may cause long-term problems by reducing the activity of the metal surfaces or clogging pores.
2. Long-term efficiency of nZVI colloidal barriers is unclear.
3. Nanoscale ZVI particles tend to aggregate, limiting their dispersability and potential reactivity.
4. Microbiological effects are likely to occur during extended application of in-ground reactive barriers.

Additional research in the following areas is needed if nZVI-mediated treatment technology is to become established:

1. Minimization of the surface passivation with respect to the contaminant degradation reaction;
2. Investigation of combined treatment with biodegradation since biodegradation by microorganisms may be important in meeting remediation targets if effluent from the iron-bearing zone contains partial degradation products that are still considered hazardous;
3. Improvement of effective treatment over longer time periods, and the determination of the efficiency of continuous removal with a cleaner surface under certain groundwater conditions;
4. An accurate means of predicting the interactions in nZVI colloidal barriers to improve the efficiency of application and to predict longevity so that reliable cost assessments can be made;
5. Understanding the chemical interactions over long time periods;
6. Understanding of both contaminant and major ion chemistry for cost-effective designs;
7. Methods of provision of oxygen in systems where this key reactant is limiting;
8. Development of efficient nZVI emplacement methods.

Decontamination of polluted groundwater is always problematic since the technological facilities used for flow through treatment systems must be designed for a high capacity and long-term operation. Thus, economic considerations, and concern about effective long-term treatment are contributing to a shift away from the established "pump and treat" treatment remediation systems toward in situ methods (Janda et al., 2004). Within this context, nZVI oxidative technologies, which do not require the use of expensive oxidants, are likely to be particularly attractive.

8
References

Agency for Toxic Substances and Disease Registry (ATSDR) (1990) *Toxicological profile for toxaphene*. US Public Health Service, US Department of Health and Human Services, Atlanta, GA.

Agency for Toxic Substances and Disease Registry (ATSDR) (1992) *Toxicological profile for chlordane*. US Public Health Service, US Department of Health and Human Services, Atlanta, GA.

Agency for Toxic Substances and Disease Registry (ATSDR) (1993) *Toxicological profile for heptachlor/heptachlor epoxide*. US Public Health Service, US Department of Health and Human Services, Atlanta, GA.

Agilar, C.; Peñalver, S.; Pocurull, E.; Borrull, F.; Marcé, R.M. (1998) Solid-phase micro-extraction and gas chromatography with mass spectrometric detection for the determination of pesticides in aqueous samples. *Journal of Chromatography A*, 795, pp. 105–115.

Agrawal, A.; Tratnyek, P.G. (1996) Reduction of nitro aromatic compounds by zero-valent iron metal. *Environmental Science and Technology*, 30(1), pp. 153–160.

Allen, D.; Shonnard, D. (2002) *Green engineering: Environmentally conscious design of chemical processes*, Prentice Hall PTR, Upper Saddle River, NJ, Chapter 1.

Al Momani, F.; Sans, C.; Esplugas, S. (2004) A comparative study of the advanced oxidation of 2,4-dichlorophenol. *Journal of Hazardous Materials*, 107(3), pp. 123–129.

Alowitz, M.J.; Scherer, M.M. (2002) Kinetics of nitrate, nitrite, and Cr(VI) reduction by iron metal. *Environmental Science and Technology*, 36, pp. 299–306.

Amaraneni, S.R. (2002) Persistence of pesticides in water, sediment and fish from fish farms in Kolleru Lake, India. *Journal of the Science of Food and Agriculture*, 82, pp. 918–923.

Andreozzi, R.; Marotta, R. (2004) Removal of benzoic acid in aqueous solution by Fe(III) homogeneous photocatalysis. *Water Research*, 38, pp. 1225–1236.

Andrews Environmental Engineering, Inc. (1994) *Use of landfarming to remediate soil contaminated by pesticides*, HWRIC TR-019 Final Report, Illinois Hazardous Waste Research and Information Center, Champaign, IL, 42pp.

Appleton, E.L. (1996) A nickel-iron wall against contaminated groundwater. *Environmental Science and Technology*, 30, pp. 536A–539A.

Arthur, E.L.; Coats, J.R. (1998) Phytoremediation. In P.C. Kearney, T. Roberts (eds.), *Pesticide Remediation in Soils and Water*, John Wiley & Sons, New York, pp. 251-283.

Augusti, R.; Dias, A.O.; Rocha, L.L.; Lago, R.M. (1998) Kinetics and mechanism of benzene derivative degradation with Fenton's reagent in aqueous medium studied by MIMS. *Journal of Physical Chemistry A*, 102, pp. 10723–10727.

Australian State of the Environment Committee (2001) *Australia: State of the Environment Report 2001*, CSIRO Publishing, Collingwood, Victoria, Australia.

Australian and New Zealand Environment and Conservation Council (1992) *Australian water quality guidelines for fresh and marine waters*. Australian and New Zealand Environment and Conservation Council, Environment Priorities and Coordination Group, Environment Australia, Canberra, Australia.

Bader, H.; Sturzenegger, V.; Hoigné, J. (1988) Photometric method for the determination of low concentrations of hydrogen peroxide by the peroxidase catalyzed oxidation of N,N-diethyl-p-phenylenediamine (DPD). *Water Research*, 2(9), pp. 1109–1115.

Balmer M.E.; Sulzberger B. (1999) Atrazine degradation in irradiated iron/oxalate systems: effects of pH and oxalate. *Environmental Science and Technology*, 33, pp. 2418–2424.

Barnabas, I.L.; Dean, J.R.; Fowlis, I.A.; Owen, S.P. (1995) Automated determination of s-triazine herbicides using solid-phase microextraction. *Journal of Chromatography A*, 705, pp. 305–312.

Barnabas, I.J.; Dean, J.R.; Hitchen, S.M.; Owen, S.P. (1994) Supercritical fluid extraction of organochlorine pesticides from an aqueous matrix. *Journal of Chromatography A*, 665(2), pp. 307–315.

Barton, D.H.R; Doller, D. (1992) The selective functionalization of saturated hydrocarbons: Gif chemistry. *Accounts of Chemical Research*, 25, pp. 504–512.

Bell, L.S.; Devlin, J.F.; Gillham, R.W.; Binning, P.J. (2003) A sequential zero-valent iron and aerobic biodegradation treatment system for nitrobenzene. *Journal of Contaminant Hydrology*, 66, pp. 201–217.

Beltran, J.; Lopez, F.J.; Cepria, O.; Hernandez, F. (1998) Solid-phase microextraction for quantitative analysis of organophosphorus pesticides in environmental water samples. *Journal of Chromatography A*, 808, pp. 257–263.

Beltran, J.; López, F.J.; Hernández, F. (2000) Solid-phase micro-extraction in pesticide residue analysis. *Journal of Chromatography A*, 885, pp. 389–404.

Benner, S.G.; Blowes, D.W.; Ptacek, C.J. (1997) A full-scale porous reactive wall for prevention of acid mine drainage. *Ground Water Monitoring and Remediation*, 17(4), pp. 99–107.

Bergendahl, J.A.; Thies, T.P. (2004) Fenton's oxidation of MTBE with zero-valent iron. *Water Research*, 38, pp. 327–334.

Bevington, P.; Robinson, D.K. (2003) Error analysis. *Data reduction and error analysis for the physical sciences*, 3rd ed., McGraw-Hill, New York, pp. 36–50.

Bielski, B.H.J.; Cabelli, D.E.; Arudi, R.L.; Ross, A.B. (1985) Reactivity of HO_2/O_2^- radicals in aqueous solution. *Journal of Physical and Chemical Reference Data*, 14(4), pp. 1041–1100.

Bielski, B.H.J.; Cabelli, D.E. (1995) Superoxide and hydroxyl radical chemistry in aqueous solution, In C.S. Foote, J.S. Valentine, A. Greenberg, J.F. Liebrnan (eds.), *Active oxygen in chemistry*, Blackie Academic & Professional, New York, p. 66.

Birke, V.; Burmeier, H.; Rosenau, D. (2003) Design, construction, and operation of tailored permeable reactive barriers. *Practice Periodical of Hazardous, Toxic, and Radioactive Waste Management*, 7(4), pp. 264–280.

Biswas, T.K.; Naismith, A.N.; Jayawardane, N.S. (2000) Performance of a land FILTER technique for pesticide removal from contaminated water. In J.A. Adams, A.K. Metherell, (eds.), *Proceedings of Soil 2000: New horizons for a new century*, Lincoln University, Canterbury Christchurch, New Zealand, pp. 23–24.

Blokhina, O.; Virolainen, E.; Fagerstedt, K.V. (1991) Antioxidants, oxidative damage and oxidation deprivation stress: A review. *Annals of Botany*, pp. 179–194.

Bogdal, M.; Lukasiewicz, J.; Pielichowski (2004) Halogenation of carbazole and other aromatic compounds with hydrohalic acids and hydrogen peroxide under microwave irradiation. *Green Chemistry*, 2, pp. 110–113.

Boronina, T.; Klabunde, K.J.; Sergeev, G. (1995) Destruction of organohalides in water using metal particles: Carbon tetrachloride/water reactions with magnesium, tin, zinc. *Environmental Science and Technology*, 29(6), pp. 1511–1517.

Bossmann, S.H.; Oliveros, E.; Göb, S.; Siegwart, S.; Dahlen, E.P.; Payawan, L., Jr.; Straub, M.; Wörner, M.; Braum, A.M. (1998) New evidence against hydroxyl radicals as reactive intermediates in the thermal and photochemically enhanced Fenton reactions. *Journal of Physical Chemistry A*, 102, pp. 5542–5550.

Bouaid, A.; Ramos, L.; Gonzalez, M.J.; Fernandez, P.; Camara, C. (2001) Solid-phase microextraction method for the determination of atrazine and four organophosphorus pesticides in soil samples by gas chromatography. *Journal of Chromatography A*, 939, pp. 13–21.

Boyd-Boland, A.A.; Magdic, S.; Pawliszyn, J.B. (1996) Simultaneous determination of 60 pesticides in water using solid-phase microextraction and gas chromatography-mass spectrometry. *Analyst*, 121(July), pp. 929–938.

Boyd-Boland, A.A.; Pawliszyn, J.B. (1996) Solid-phase microextraction coupled with high-performance liquid chromatography for the determination of alkylphenol ethoxylate surfactants in water. *Analytical Chemistry*, 68, pp. 1521–1529.

Bozec, N.L.; Compère, C.; L'Her, M.; Laouenan, A.; Costa, D.; Marcus, P. (2001) Influence of stainless steel surface treatment on the oxygen reduction reaction in seawater. *Corrosion Science*, 43, pp. 765–786.

Brausam, A.; van Eldik, R. (2004) Further mechanistic information on the reaction between $Fe^{III}EDTA$ and hydrogen peroxide: Observation of a second reaction step and importance of pH. *Inorganic Chemistry*, 43, pp. 5351–5359.

Brooks, A.; Westhorpe, D.; Bales, M. (1996) *The impacts of pesticides on riverine environments*. Department of Land and Water Conservation, Technical Services Directorate, Sydney, N.S.W.

Bucknall, T.; Edwards, H.E. (1978) The formation of malonaldehyde in irradiated carbohydrates. *Carbohydrate Research*, 62, pp. 49–59.

Buchholz, K.D.; Pawliszyn, J. (1993) Determination of phenols by solid-phase microextraction and gas chromatographic analysis. *Environmental Science and Technology*, 27, pp. 2844–2848.

Buesseler, K.O.; Boyd, P.W. (2003) Will ocean fertilization work? *Science*, 300, p. 67–68.

Buettner, G.R.; Doherty, T.P.; Patterson, L.K. (1983) The kinetics of the reaction of superoxide radical with Fe(III) complexes of EDTA, DETAPAC, and HEDTA. *FEBS Letters*, 158, p. 143.

Bull, C.; McClune, G.J.; Fee, J.A. (1983) The Mechanism of Fe-EDTA catalyzed superoxide dismutation. *Journal of the American Chemical Society*, 105, pp. 5290–5300.

Burris, D.R.; Campbell, T.J.; Manoranjan, V.S. (1995) Sorption of trichloroethylene and tetrachloroethylene in a batch reactive metallic iron-water system. *Environmental Science and Technology*, 29, pp. 2850–2855.

Buxton, G.V.; Greenstock, C.L.; Helman, W.P.; Ross, A.B. (1988) Critical review of rate constants for reactions of hydrated electrons, hydrogen atoms and hydroxyl radicals in aqueous solution. *Journal of Physical and Chemical Reference Data*, 17, pp. 513–886.

Calvo, E.J.; Schiffrin, D.J. (1984) The reduction of hydrogen peroxide on passive iron in alkaline solutions. *Journal of Electroanalytical Chemistry*, 163, pp. 257–275.

Cantrell, K.J.; Kaplan, D.I. (1997) Zero-valent iron colloid emplacement in sand columns. *Journal of Environmental Engineering*, 123, pp. 499–505.

Cao, J.; Wei, L.; Huang, Q.; Wang, L.; Han, S. (1999) Reducing degradation of azo dyes by zero-valent iron in aqueous solution. *Chemosphere*, 38, pp. 565–571.

Capangpangan, M.B.; Suffet, I.H. (1996) Optimization of a drying method for filtered suspended solids from natural waters for supercritical fluid extraction analysis of hydrophobic organic compounds. *Journal of Chromatography A*, 738, pp. 253–264.

Capangpangan, M.B.; Suffet, I.H. (1997) Validation studies for a new supercritical fluid extraction method for the isolation of hydrophobic organic compounds from filtered suspended solids. *Journal of Chromatography A*, 782, pp. 247–256.

Cerejeira, M.J.; Viana, P.; Batista, S.; Pereira, T.; Silva, E.; Valério, M.J.; Silva, A.; Ferreira, M.; Silva-Fernandes, A.M. (2003) Pesticides in Portuguese surface and ground waters. *Water Research*, 37, pp. 1055–1063.

Charizopoulos, E.; Papadopoulou-Mourkidou, E. (1999) Occurrence of pesticides in rain of the Axios river basin, Greece. *Environmental Science and Technology*, 33, pp. 2363–2368.

Charlet, L.; Liger, E.; Gerasimo, P. (1998) Decontamination of TCE- and U-rich waters by granular iron: Role of sorbed Fe(II). *Journal of Environmental Engineering*, 124(1), pp. 25–30.

Chen, J.L.; Al-Abed, S.R.; Ryan, J.A.; Li, Z. (2001) Effects of pH on dechlorination of trichloroethylene by zero-valent iron. *Journal of Hazardous Materials*, 83(3), pp. 243–254.

Chen, R. Z.; Pignatello, J. J. (1997) Role of quinone intermediates as electron shuttles in Fenton and photoassisted Fenton oxidations of aromatic compounds. *Environmental Science and Technology*, 31(8), pp. 2399–2406.

Cheng, I.F.; Breen, K. (2000) The ability of four flavanoids, baicilien, luteolin, naringenin and quercetin, to suppress the Fenton reaction of the iron-ATP complex. *Biometals.* 13, pp. 77–83.

Cheng, I.F.; Fernando, Q.; Korte, N. (1997), Electrochemical dechlorination of 4-chlorophenol to phenol. *Environmental Science and Technology*, 31, pp. 1074–1078.

Cheng, Z.; Li, Y.; Chang, W. (2003) Kinetic deoxyribose degradation assay and its application in assessing the antioxidant activities of phenolic compounds in a Fenton-type reaction system. *Analytica Chimica Acta*, 478, pp. 129–137.

Chew, C.F.; Zhang, T.C. (1999) Abiotic degradation of nitrites using zero-valent iron and electrokinetic processes. *Environmental Engineering Science*, 16(5), pp. 389–401.

Chiron, S.; Fernandez-Alba, A.; Rodriguez, A.; Garcia-Calvo, E. (2000) Pesticide chemical oxidation: State-of-the-art. *Water Research*, 34, pp. 366–377.

Cochran, R.C.; Kishiyama, J.; Aldous, C.; Carr, W.C., Jr. (1995) Chlorpyrifos: Hazard assessment based on a review of the effects of short-term and long-term exposure in animals and humans. *Food and Chemical Toxicology*, 33(2), pp. 165–172.

Choe, S.; Chang, Y.-Y.; Hwang, K.-Y.; Khim, J. (2000) Kinetics of reductive denitrification by nanoscale zero-valent iron. *Chemosphere*, 41, pp. 1307–1311.

Choe, S.; Lee, S.-H.; Chang, Y.-Y.; Hwang, K.-Y.; Khim, J. (2001) Rapid reductive destruction of hazardous organic compounds by nanoscale Fe⁰. *Chemosphere*, 42, pp. 367–372.

Collins, T. (1994) Designing ligands for oxidizing complexes. *Accounts of Chemical Research*, 27, pp. 279–285.

Collins, T. (2002) TAML oxidant activators: A new approach to activation of hydrogen peroxide for environmentally significant problems. *Accounts of Chemical Research*, 35, pp. 782–790.

Coupe, R.H.; Thurman, E.M.; Zimmerman, L.R. (1998) Relation of usage to the occurrence of cotton and rice herbicides in three streams of the Mississippi Delta. *Environmental Science and Technology*, 32, pp. 3673–3680.

Crepeau, K.L.; Kuivila, K.M. (2000) Rice pesticide concentrations in the Colusa basin drain and the Sacramento River, California, 1990–1993. *Journal of Environmental Quality*, 29, pp. 926–935.

Croft, S.; Gilbert, B.C.; Smith, L.; Whitwood, A.C. (1992) An ESR investigation of the reactive intermediate generated in the reaction between Fe^{II} and H_2O_2 in aqueous solution. *Free Radical Research Communications*, 17(1), pp. 22–39.

Crow, D.R. (1984) *Principles and applications of electrochemistry*, Chapman and Hall, New York.

Das, S.; Schuchmann, M.N.; Schuchmann, H.P.; Vonsonntag, C. (1987) The production of the superoxide radical-anion by the OH radical-induced oxidation of trimethylamine in oxygenated aqueous-solution—The kinetics of the hydrolysis of dimethylamine. *Chemische Berichte-Recueil*, 120(3), pp. 319–323.

Davenport, A.J.; Oblonsky, L.J.; Ryan, M.P.; Toney, M.F. (2000) The structure of the passive film that forms on iron in aqueous environments. *Journal of the Electrochemical Society*, 147, pp. 2162–2173.

Day, S. R.; O'Hannesin, S. F.; Marsden, L. (1999) Geotechnical techniques for the construction of reactive barriers. *Journal of Hazardous Materials*, 67, pp. 285–297.

Devlin, J.F.; Klausen, J.; Schwarzenbach, R.P. (1998) Kinetics of nitroaromatic reduction on granular iron in recirculating batch experiments. *Environmental Science and Technology*, 32, pp. 1941–1947.

Dombek, T.; Dolan, E.; Schultz, J.; Klarup, D. (2001) Rapid reductive dechlorination of atrazine by zero-valent iron under acidic conditions. *Environmental Pollution*, 111, pp. 21–27.

Donaldson, D.; Kiely, T.; Grube, A. (2002) *Pesticides industry sales and usage*, US Environment Protection Agency, Washington, D.C.

Dorfman, L.M.; Buhler, R.E.; Taub, I.A. (1962) Pulse radiolysis studies. I. Transient spectra and reaction-rate constants in irradiated aqueous solutions of benzene. *Journal of Chemical Physics*, 36(11), pp. 3051–3061.

Dowling, K.C.; Lemley, A.T. (1995) Organophosphate insecticide degradation by nonamended and cupric-amended Fenton's reagent in aqueous solution. *Journal of Environmental Science and Health B*, 30(5), pp. 585–604.

Dugay, J.; Miege, C.; Hennion, M.-C. (1998) Effect of the various parameters governing solid-phase microextraction for the trace-determination of pesticides in water. *Journal of Chromatography A*, 795, pp. 27–42.

Eberhardt, M.K.; Yoshida, M. (1973) Radiation-induced hemolytic aromatic substitution. I. Hydroxylation of nitrobenzene, chlorobenzene, and toluene. *Journal of Physical Chemistry*, 77(5), pp. 589–597.

Eisert, R.; Levsen, K. (1995) Determination of pesticides in aqueous samples by solid-phase microextraction in-line coupled to gas chromatography-mass spectrometry. *Journal of the American Society for Mass Spectrometry*, 6, pp. 1119–1130.

Elliott, D.W.; Zhang, W.-X. (2001) Field assessment of nanoscale bimetallic particles for groundwater treatment. *Environmental Science and Technology*, 35, pp. 4922–4926.

Englemann, M.; Bobier, R.; Hiatt T.; Cheng, F. (2003) Variablity of the Fenton reaction characteristics of EDTA, DTPA and citrate complexes of iron, *Biometals*, 16, p. 519.

Eykholt, G.; Davenport, D.T. (1998) Dechlorination of the chloroacetanilide herbicides alachlor and metolachlor by iron metal. *Environmental Science and Technology*, 32, pp. 1482–1487.

Fee, J.A.; DiCorleto, P.E. (1973) Observation on the oxidation-reduction properties of bovine erythrocyte superoxide dismutase. *Biochemistry*, 12, pp. 4893–4899.

Feig, A.; Becker, M.; Schindler, S.; van Eldik, R.; Lippard, S. (1996) Mechanistic studies of the formation and decay of diiron (III) peroxo complexes in the reaction of diiron(II) precursors with dioxygen. *Inorganic Chemistry*, 35, pp. 2590–2601.

Feig, A.; Lippard, S. (1994) Reactions of non-heme iron(II) centers with dioxygen in biology and chemistry, *Chemical Review*, 94, pp. 759–805.

Feitz, A.J.; Waite, T.D. (2003) Kinetic modeling of TiO_2-catalyzed photodegradation of trace levels of microcystin-LR. *Environmental Science and Technology*, 37, pp. 561–568.

Feitz, A.J.; Waite, T.D.; Joo, S.H. (2005) Method for decontaminating surfaces. International Patent, 29 pp. WO 2005053797.

Feitz, A.J.; Waite, T.D.; Joo, S.H.; Guan, J.; Biswas, T.K. (2002) *Development of nanosized zero-valent iron particles for organic contaminant degradation*. CRC WMPC Report, Centre for Water and Waste Technology, UNSW, Sydney, Australia.

Felsot, A.S.; Racke, K.D.; Hamilton, D.J. (2003) Disposal and degradation of pesticide waste. *Reviews of Environmental Contamination and Toxicology*, 177, pp.123–200.

Fennelly, J.P.; Roberts, A.L. (1998) Reaction of 1,1,1-trichloroethane with zero-valent metals and bimetallic reductants. *Environmental Science and Technology*, 32, pp. 1980–1988.

Fenton, H.J.H (1894) Oxidation of tartaric acid in presence of iron, *Journal of Chemical Society*, 65, p. 899.

Ferrano, S.P.; Lee, H.; Smith, L.M.; Ozretich, R.J.; Specht, D.T. (1991) Accumulation factors for eleven polychlorinated biphenyl congeners. *Bulletin of Environmental Contamination and Toxicology*, 46(2), pp. 276–283.

Fruchter, J. S.; Cole, C. R.; Williams, M. D.; Vermeul, V. R.; Amonette, J. E.; Szecsody, J. E.; Istok, J. D.; Humphrey, M. D. (2000) Creation of a subsurface permeable treatment zone for aqueous chromate contamination using in situ redox manipulation. *Ground Water Monitoring and Remediation*, 20(2), pp. 66–67.

Ghauch, A. (2001) Degradation of benomyl, picloram, and dicamba in a conical apparatus by zero-valent iron powder. *Chemosphere*, 43, pp. 1109–1117.

Ghauch, A.; Gallet, C.; Charef, A.; Rima, J.; Martin-Bouyer, M. (2001) Reductive degradation of carbaryl in water by zero-valent iron. *Chemosphere*, 42, pp. 419–424.

Ghauch, A.; Suptil, J. (2000) Remediation of s-triazines contaminated water in a laboratory scale apparatus using zero-valent iron powder. *Chemosphere*, 41, pp. 1835–1843.

Gillham, R.W.; O'Hannesin, S.F. (1994) Enhanced degradation of halogenated aliphatics by zero-valent iron. *Ground Water*, 32(6), pp. 958–967.

Green, J.M. (1996) A practical guide to analytical method validation. *Analytical Chemistry*, 68(9), pp. A305–A309.

Guerin, T.F.; Kimber, S.W.L.; Kennedy, I.R. (1992) Efficient one-step method for the extraction of cyclodiene pesticides from aqueous media and the analysis of their metabolites. *Journal of Agricultural and Food Chemistry*, 40, pp. 2309–2314.

Gunier, R.B.; Harnly, M.E.; Reynolds, P.; Hertz, A.; von Behren, J. (2001) Agricultural pesticide use in California: pesticide prioritization, use densities, and population distributions for a childhood cancer study. *Environmental Health Perspectives*, 109, pp. 1071–1078.

Gupta, P.K.; Salunkhe, D.K. (1985) *Modern toxicology, Vol III: Immuni and clinical toxicology*, Metropolitan, New Delhi.

Gupta, S.; Stadler, M.; Noser, C.; Ghosh, A.; Steinhoff, B.; Dieter, L.; Horwitz, C.; Schramm, K.; Collins, T. (2002) Rapid total destruction of chlorophenols by activated hydrogen peroxide. *Science*, 296, pp. 326–328.

Gutteridge, J.M. (1987) Ferrous-salt-promoted damage to deoxyribose and benzoate. The increased effectiveness of hydroxyl-radical scavengers in the presence of EDTA. *Biochemical Journal*, 243, pp. 709–714.

Halliwell, B.; Gutteridge, J.; Aruoma, O. (1987) The deoxyribose method: A simple "Test-Tube" assay for determination of rate constants for reactions of hydroxyl radicals. *Analytical Chemistry*, 165, pp. 215–219.

Handshaw, N.; Martell, A. (1998) Ferrous chelates of EDTA, HEDTA and SHBED, *Inorganic Chemistry*, 27, pp. 1297–1298.

Harris, C.R.; Kennedy, I.R. (1996) *Pesticides in perspective*, Australian Cotton Grower, Sydney.

Hartley, D.; Kidd, H. (1983) *The agrochemicals handbook*, Royal Society of Chemistry, London, U.K.

Hayon, E.; Ibata, T.; Lichtin, N.N.; Simic, M. (1970) Sites of attack of hydroxyl radicals on amides in aqueous solution. *Journal of American Chemical Society*, 92(13), pp. 3903–3998.

Hernandez, F.; Beltran, J.; Lopez, F.J.; Gaspar, J.V. (2000) Use of solid-phase microextraction for the quantitative determination of herbicides in soil and water samples. *Analytical Chemistry*, 72, pp. 2313–2322.

Hozalski, R.M.; Zhang, L.; Arnold, W.A. (2001) Reduction of haloacetic acids by Fe^0: Implications for treatment and fate. *Environmental Science and Technology*, 35, pp. 2258–2263.

Hsieh, Y.N.; Liu, L.F.; Wang, Y.S. (1998) Uptake, translocation and metabolism of the herbicide molinate in tobacco and rice. *Pesticide Science*, 53(2), pp. 149–154.

Hundal, L.S.; Singh, J.; Bier, E.L.; Shea, P.J.; Comfort, S.D.; Powers, W.L. (1997) Removal of TNT and RDX from water and soil using iron metal. *Environmental Pollution*, 97, pp. 55–64.

Hung, D.Q.; Thiemann, W. (2002) Contamination by selected chlorinated pesticides in surface waters in Hanoi, Vietnam. *Chemosphere*, 47, pp. 357–367.

Hutson, D.H.; Roberts,T.R. (1990) *Environmental fate of pesticides*, John Wiley & Sons, Chichester , England.

Indian Pesticides Industry Scope Marketing & Information Solutions (IPISMIS) (2001) Indian pesticides industry. http://www.researchmarkets.com. Accessed in August 2004.

International Program on Chemical Safety (INCHEM) (1989) Environmental health criteria 91. http://www.inchem.org. Accessed in August 2004.

Ishikawa, T.; Kondo, Y.; Yasukawa, A.; Kandori, K. (1998) Formation of magnetite in the presence of ferric oxyhydroxides. *Corrosion Science*, 40(7), pp. 1239–1251.

Janda, V.; Vasek, P.; Bizova, J.; Belohlav, Z. (2004) Kinetic models for volatile chlorinated hydrocarbons removal by zero-valent iron. *Chemosphere*, 54, pp. 917–925.

Jayawardane, N.S.; Biswas, T.K.; Blackwell, J.; Cook, F.J. (2001) Management of salinity and sodicity in a land FILTER system for treating saline wastewater on a saline-sodic soil. *Australian Journal of Soil Research*, 39, pp. 1247–1258.

Joo, S.H.; Feitz, A.J.; Waite, T.D. (2002a) Agrochemical degradation using nano-scale zero valent iron, ZVI/H_2O_2, and Fenton's reagent. In *Proceedings of the Second International*

Conference on oxidation and reduction technologies for in-situ treatment of soil and groundwater, 17–21 November, Toronto, Ontario, Canada.

Joo, S.H.; Feitz, A.J.; Sedlak, D.L.; Waite, T.D. (2005) Quantification of the oxidizing capacity of nanoparticulate zero-valent iron, *Environmental Science and Technology*, 39(5, special issue), pp. 1263–1268.

Joo, S.H.; Feitz, A.J.; Waite, T.D. (2004) Oxidative degradation of the carbothioate herbicide, molinate, using nanoscale zero-valent iron. *Environmental Science and Technology*, 38, pp. 2242–2247.

Jovancicevic, V.; Bockris, J.O'M. (1986) The mechanism of oxygen reduction on iron in neutral solutions. *Journal of Electrochemical Society: Electrochemical Science and Technology*, 133(9), pp. 1797–1807.

Kamolpornwijit, W.; Liang, L.; West, O.R.; Moline, G.R.; Sullivan, A.B. (2003) Preferential flow path development and its influence on long-term PRB performance: Column study. *Journal of Contaminant Hydrology*, 66(3–4), pp. 161–178.

Kanatharana, P.; Chindarasamee, C.; Kaewnarong, B. (1993) Extracting solvents for pesticide residues. *Journal of Environmental Science and Health part A—Environmental Science and Engineering & Toxic and Hazardous Substance Control*, 28(10), pp. 2323–2332.

Kanel, S.R.; Manning, B.; Charlet, L.; Choi, H. (2005) Removal of arsenic (III) from groundwater by nanoscale zero-valent iron. *Environmental Science and Technology*, 39(5), pp. 1291–1298.

Kaplan, D.I.; Catrell, K.J.; Wietsma, T.W.; Potter, M.A. (1996) Retention of zero-valent iron colloids by sand columns: Application to chemical barrier formation. *Journal of Environmental Quality*, 25, pp. 1086–1094.

Keene J.P. (1964) Pulse radiolysis of ferrous sulfate solution. *Radiation Research*, 22(1), p. 14.

Kelly, K.; Reed, N. (1996) *Pesticides for evaluation as candidate toxic air contaminants*,California Department of Pesticide Regulation, Sacramento, CA.

Kentucky Department of Agriculture Richie Farmer, Commissioner (KDARFC) (2004) *Guidelines for atrazine use and application for groundwater and surface water protection: Best management practices.* University of Kentucky College of Agriculture, Division of Conservation, and US Natural Resources Conservation Service Western Kentucky University, Kentucky State University, Syngenta, BMP-6.

Kim, Y.-H.; Carraway, E.R. (2000) Dechlorination of pentachlorophenol by zero-valent iron and modified zero-valent irons. *Environmental Science and Technology*, 345, pp. 2014–2017.

Kitajima, N.; Fukuzumi, S.; Ono, Y. (1978) Formation of superoxide ion during decomposition of hydrogen peroxide on supported metal oxides. *Journal of Physical Chemistry*, 82(13), pp. 1505–1509.

Klein, G.W.; Bhatla, K.; Madhavan, V.; Schuler, R.H. (1975) Reaction of hydroxyl radicals with benzoic acid: Isomer distribution in the radical intermediates. *Journal of Physical Chemistry*, 79, pp. 1767–1774.

Konstantinou, I.K.; Sakkas, V.A.; Albanis, T.A. (2001) Photocatalytic degradation of the herbicides propanil and molinate over aqueous TiO_2 suspensions: Identification of intermediates and the reaction pathway. *Applied Catalysis B: Environmental* 34, pp. 227–239.

Konstantinou, I.K.; Sakellarides, T.M.; Sakkas, V.A.; Albanis, T.A. (2001) Photocatalytic degradation of selected *s*-triazine herbicides and organophosphorus insecticides over aqueous TiO_2 suspensions. *Environmental Science and Technology*, 35, pp. 398–405.

Kormann, C.; Bahnemann, D.W.; Hoffmann, M.R. (1991) Photolysis of chloroform and other organic molecules in aqueous titanium dioxide suspensions. *Environmental Science and Technology*, 25, pp. 494–500.

Kremer, M.L. (1999) Mechanism of the Fenton reaction: Evidence for a new intermediate. *Physical Chemistry Chemical Physics*, 1, pp. 3595–3605.

Lackovic, J.A.; Nikolaidis, N.P.; Dobbs, G.M. (2000) Inorganic arsenic removal by zero-valent iron. *Environmental Engineering Science*, 17, pp. 29–39.

Lancaster, M. (2002) *Green chemistry: An introductory text*, Royal Society of Chemistry: Cambridge.

Larson, R.; Weber, E. (1994) *Reactions mechanisms in environmental organic chemistry*, CRC Press, Boca Raton, FL, pp. 315–341.

Lee, S.; MaLaughlin, R.; Harnly, M.; Gunier, R.; Kreutzer, R. (2002) Community exposures to airborne agricultural pesticides in California: Ranking of inhalation risks. *Environmental Health Perspectives*, 110, pp. 1175–1184.

Legrini, O.; Oliveros, E.; Braun, A.M. (1993) Photochemical processes for water treatment. *Chemical Reviews*, 93, pp. 671–698.

Li, Tie; Farrell, J. (2000) Reductive dechlorination of trichloroethene and carbon tetrachloride using iron and palladized-iron cathodes. *Environmental Science and Technology*, 34, pp. 173–179.

Li, Y.; Zhang, J. (1999) Agricultural diffuse pollution from fertilizers and pesticides in China. *Water Science and Technology*, 39(3), pp. 25–32.

Liang, L.; Korte, N.; Gu, B.; Puls, R.; Reeter, C. (2000) Geochemical and microbial reactions affecting the long-term performance of in situ "iron barriers." *Journal of Advances in Environmental Research*, 4, pp. 273–286.

Liang, L.; Sullivan, A.B.; West, O.R.; Moline, G.R.; Komolpoinwijik, W. (2003) Predicting the precipitation of mineral phases in permeable reactive barriers. *Environmental Engineering Science*, 20(6), pp. 635–653.

Lin, S.; Gurol, M.D. (1998) Catalytic decomposition of hydrogen peroxide on iron oxide: Kinetics, mechanism, and implications. *Environmental Science and Technology*, 32, pp. 1417–1423.

Lopez-Alcala, J.M.; Puerta-Vizcaino, M.C.; Gonzalez-Vilchez, F. (1984) A redetermination of sodium aqua(edta)ferrate(III) dihydrate. *Acta Cyrstallographica*, C40, pp. 939–941.

Lopez-Avila, V.; Young, R.; Tehrani, J.; Damian, J.; Hawthorne, S.; Dankers, J.; Heiden, C.V.D. (1994) Mini-round-robin study of a supercritical fluid extraction method for polynuclear aromatic hydrocarbons in soils with dichloromethane as a static modifier. *Journal of Chromatography A*, 672(1–2), pp. 167–175.

Los Alamos National Laboratory (LANL) (2005) Iron: Biological role. http://periodic.lanl.gov/elements/26.html. Accessed in June 2005.

Lücking, F.; Köser, H.; Ritter, A. (1998) Iron powder, graphite and activated carbon as catalysts for the oxidation of 4-chlorophenol with hydrogen peroxide in aqueous solution. *Water Research*, 32(9), pp. 2607–2614.

Macounová, K.; Krýsová, H.; Ludvic, J.; Jirkovský, J. (2003) Kinetics of photocatalytic degradation of diuron in aqueous colloidal solutions of Q-TiO$_2$ particles. *Journal of Photochemistry and Photobiology A: Chemistry*, 156, pp. 273–282.

Magdic, S.; Boyd-Boland, A.; Jinno, K.; Pawliszyn, J.B. (1996) Analysis of organophosphorus insecticides. *Journal of Chromatography A*, 736, pp. 219–228.

Magdic, S.; Pawliszyn, J.B. (1996) Analysis of organochlorine pesticides using solid-phase micro-extraction. *Journal of Chromatography A*, 723, pp. 111–122.

Mall, I.D.; Raju, K.; Kumar, N. (2003) Pesticide industry: Impact on environment. *Chemical Engineering World*, 38(1), pp. 116–118, 121–123.

Matheson, L.J.; Tratnyek, P.G. (1994) Reductive dehalogenation of chlorinated methanes by iron metal. *Environmental Science and Technology*, 28(12), pp. 2045–2053.

Matin, M.A.; Malek, M.A.; Amin, M.R.; Rahman, S.; Khatoon, J.; Rahman, M.; Aminuddin, M.; Mian, A.J. (1998) Organochlorine insecticide residues in surface and underground water regions of Bangladesh. *Agriculture, Ecosystems and Environment*, 69, pp. 11–15.

Matisová, E.; Medved'ová, M.; Vraniaková, J.; Šimon, P. (2002) Optimisation of solid-phase microextraction of volatiles. *Journal of Chromatography A*, 960, pp. 159–164.

Mazellier, P.; Jirkovsky, J.; Bolte, M. (1997) Degradation of diuron photoinduced by iron (III) in aqueous solution. *Pesticide Science*, 49, pp. 259–267.

Mazellier, P.; Sulzberger, B. (2001) Diuron degradation in irradiated, heterogeneous iron/oxalate systems: The rate-determining step. *Environmental Science and Technology*, 35, pp. 3314–3320.

McKim, J.M. (1994) Physiological and biological mechanisms that regulate the accumulation and toxicity of environmental chemicals in fish. In J.L. Hamelink, P.F. Landrum, H.L. Bergman, W.H. Benson, (eds.), *Bioavailability: physical, chemical and biological interactions*. Lewis, Publishers, Boca Raton, FL.

McKinzi, A.M.; Dichristina, T.J. (1999) Microbially driven Fenton reaction for transformation of pentachlorophenol. *Environmental Science and Technology*, 33, pp. 1886–1891.

Meunier, B. (2002) Catalytic degradation of chlorinated phenols. *Science*, 296, pp. 270–271.

Mill, T.; Haag, W.R. (1989) Novel metal/peroxide systems for the treatment of organic compounds in drinking water. Preprint Extended Abstract presented before the Division of Environmental Chemistry, American Chemical Society, Miami Beach, FL, September 10–15.

Mindbranch (2001) Indian pesticides industry market report. http:// www.mindbranch.com. Accessed in August 2004.

Mishra, R.; Shukla, S.P. (1997) Impact of endosulfan on lactate dehydrogenase from the freshwater catfish clarias batrachus. *Pesticide Biochemistry and Physiology*, 57, pp. 220–234.

Mizuta, T.; Wang, J.; Katsuhiko, M. (1995) Molecular structures of Fe(II) complexes with mono- and di-protonated ethylenediamine-N,N,N',N'-tetraacetate (Hedta and H$_2$edta), as determined by X-ray crystal analyses. *Inorganica Chimica Acta*, 230, pp. 119–125.

Mizuta, T.; Wang, J.; Miyoshi, K. (1993) A seven-coordinate structure of iron(II)-EDTA complex as determined by X-ray crystal analysis. *Bulletin of the Chemical Society of Japan*, 66, pp. 2547–2551.

Mizuta, T.; Yamamoto, T.; Katsuhiko, M.; Yoshihiko, K. (1990) The ligand field stabilization effect of the metal-ligand bond distances in octahedral metal complexes with edta-type ligands. *Inorganica Chimica Acta*, 175, pp. 121–126.

Morrison, S.J.; Metzler, D.R.; Dwyer, B.P. (2002) Removal of As, Mn, Mo, Se, U, V and Zn from groundwater by zero-valent iron in a passive treatment cell: Reaction progress modeling. *Journal of Contaminant Hydrology*, 56(1–2), pp. 99–116.

Muftikian, R.; Nebesny, K.; Fernando, Q.; Korte, N. (1996) X-ray photoelectron spectra of the palladium-iron bimetallic surface used for the rapid chlorination of chlorinated organic environmental contaminants. *Environmental Science and Technology*, 30, pp. 3593–3596.

Muszkat, L. (1998) Photochemical processes. In P.C. Kearney, T. Roberts, (eds.), *Pesticide remediation in soils and water*, John Wiley & Sons, New York, pp. 307–337.

Nadezhdin, A. D.; Kozlov, Y. N.; Purmalis, A. (1976) Iron(3+) ion-catalyzed decomposition of hydrogen peroxide in the presence of tetranitromethane: Determination of the chain prolongation reaction rate constant. *Zhurnal Fizicheskoi Khimii*, 50, pp. 910–912.

Nam, S.; Tratnyek, P.G. (2000) Reduction of azo dyes with zero-valent iron. *Water Research*, 34, pp. 1837–1845.

Naqvi, S.M.; Vaishnavi, C. (1993) Bioaccumulative potential and toxicity of endosulfan insecticide to non-target animals: A mini review. *Comparative Biochemistry and Physiology*, 105C, p. 347.

Natangelo, M.; Tavazzi, S.; Benfenati, E. (2002) Evaluation of solid phase microextraction-gas chromatography in the analysis of some pesticides with different mass spectrometric techniques: Application to environmental waters and food samples. *Analytical Letters*, 35(2), pp. 327–338.

National Center for Toxic and Persistent Substances (1995) *Pesticides in the aquatic environment*. Report of the National Rivers Authority, Water Quality Series No. 26, October, HMSO, London.

Naval Facilities Engineering Service Center (NFESC) (2004) Permeable reactive barrier: Remediation of chlorinated solvents in groundwater. http://enviro.nfesc.navy/. Accessed in August 2004.

Neese, F.; Solomon, E. (1998) Detailed spectroscopic and theoretical studies on $[Fe(EDTA)(O_2)]^{3-}$: Electronic structure of the side-on ferric-peroxide bond and its relevance to reactivity. *Journal of the American Chemical Society*, 120, pp. 12829–12848.

Nerín, C.; Batlle, R.; Sartaguda, M.; Pedrocchi, C. (2002) Supercritical fluid extraction of organochlorine pesticides and some metabolites in frogs from National Park of Ordesa and Monte Perdido. *Analytica Chimica Acta*, 464, pp. 303–312.

Nohara, S.; Hanazato, T.; Iwakuma, T. (1997) Pesticide residue flux from rainwater into Lake Nakanuma in the rainy season. *Rikusuigaku Zasshi*, 58(4), pp. 385–393.

Noordkamp, E.R.; Grotenhuis, J.T.C.; Rulkens, W.H. (1997) Selection of an efficient extraction method for the determination of polycyclic aromatic hydrocarbons in contaminated soil and sediment. *Chemosphere*, 35(9), pp. 1907–1917.

Noradoun, C.E.; Cheng, I.F. (2005) EDTA Degradation induced by oxygen activation in a zerovalent iron/air/water system. *Environmental Science and Technology*, 39(18), pp. 7158–7163.

Noradoun, C.; Engelmann, M.D.; McLaughlin, M.; Hutcheson, R.; Breen, K.; Paszczynski, A.; Cheng, I.F. (2003) Destruction of chlorinated phenols by dioxygen activation under aqueous room temperature and pressure conditions. *Industrial and Engineering Chemistry Research*, 42, pp. 5024–5030.

Noradoun, C.E.; Mekmaysy, C.S.; Hutcheson, R.M.; Cheng, I.F. (2005) Detoxification of malathion a chemical warfare agent analog using oxygen activation at room temperature and pressure. *Green Chemistry*, 7(6), pp. 426–430.

Nowack, B. (2002) Environmental Chemistry of Aminopolycarboxylate Chelating Agents. *Environmental Science and Technology*, 36, pp. 4009–4016.

Nowell, L.H.; Capel, P.D.; Dileanis, P.D. (1999) *Pesticides in stream sediment and aquatic biota: distribution, trends, and governing factors*, Lewis Publishers, Boca Raton, FL.

NRA (2000) *The NRA review of diazinon*. National Registration Authority for Agricultural and Veterinary Chemicals, Canberra, Australia.

Odziemkowski, M.S. (2000) The role of oxide films in the reduction of *n*-nitrosodimethylamine with reference to the iron groundwater remediation technology. *Oxide films proceedings of the international symposium*, Vol. 2000-4, The Electrochemical Society, Pennington, NJ, pp. 357–368.

O'Hannesin, S.F.; Gillham, R.W. (1998) Long-term performance of an in situ "iron wall" for remediation of VOCs. *Groundwater*, 36(1), pp. 164–170.

Oliver, B.G. (1985) Bioconcentration factors of some halogenated organics for rainbow trout: Limitations in their use for predictions of environmental residues. *Environmental Science and Technology*, 19, pp. 842–849.

Oros, D.R.; Jarman, W.M.; Lowe, T.; David, N.; Lowe, S.; David, J.A. (2003) Surveillance for previously unmonitored organic contaminants in the San Francisco Estuary. *Marine Pollution Bulletin*, 46, pp. 1102–1110.

Orth, W.S.; Gillham, R.W. (1996) Dechlorination of trichloroethene in aqueous solution using Fe-O. *Environmental Science and Technology*, 30, pp. 66–71.

Paldy, A.; Puskar, N.; Farkas, I. (1988) Pesticide use related to cancer incidence as studied in a rural district of Hungary. *Science Total Environment*, 73, pp. 224–229.

Paune, F.; Caixach, J.; Espadaler, I.; Om, J.; Rivera, J. (1998) Assessment on the removal of organic chemicals from raw and drinking water at a Llobregat River water works plant using GAC. *Water Research*, 32, pp. 3313–3324.

Pera-Titus, M.; Garcia-Molina, V.; Banos, M.A.; Gimenez, J.; Esplugas, S. (2004) Degradation of chlorophenols by means of advanced oxidation processes: A general review. *Applied Catalysis B: Environmental*, 47(4), pp. 219–256.

Pesticide Management Education Program (PMEP) (2003) The pesticide management education program at Cornell University. http://pmep.cce.cornell.edu/. Assessed in August 2004.

Pichat, P. (1997) Photocatalytic degradation of aromatic and alicyclic pollutants in water: By-products, pathways and mechanisms. *Water Science and Technology*, 35(4), pp. 73–78.

Pourbaix, M. (1966) *Atlas of electrochemical equilibria in aqueous solutions*, Pergamon, Oxford, p. 644.

Powell, R.M.; Puls, R.W. (1997) Proton generation by dissolution of intrinsic or augmented aluminosilicate minerals for in situ contaminant remediation by zero-valence-state iron. *Environmental Science and Technology*, 31, pp. 2244–2251.

Powell, R.M.; Puls, R.W.; Blowes, D.W.; Vogan, J.L.; Gillham, R.W.; Powell, P.D.; Schultz, D.; Landis, R.; Sivavic, T. (1998) *Permeable reactive barrier technologies for contaminated remediation*. US Environmental Protection Agency, EPA/600/R-98/125, Ada, OK.

Powell, R.M.; Puls, R.W.; Hightower, S.K.; Sabatini, D.A. (1995) Coupled iron corrosion and chromate reduction: mechanisms for subsurface remediation. *Environmental Science and Technology*, 29, pp. 1913–1922.

Powell, R.M.; Puls, R.W.; Powell, P.O. (2002) Remediation of chlorinated and recalcitrant compounds. In A.R. Gavaskar, A.S.C. Chen (eds.), *Proceedings of the Third International Conference on Remediation of Chlorinated and Recalcitrant Compounds*, Monterey, CA, May 20–23, pp. 104–111, Battelle Press, Columbus, OH.

Pratap, K.; Lemley, A.T. (1998) Fenton electrochemical treatment of aqueous atrazine and metolachlor. *Journal of Agricultural and Food Chemistry*, 46, pp. 3285–3291.

Pratt, A.R.; Blowes, D.W.; Ptacek, C.J. (1997) Products of chromate reduction on proposed subsurface remediation material. *Environmental Science and Technology*, 31, pp. 2492–2498.

Psillakis, E; Goula, G.; Kalogerakis, N; Mantzavinos, D. (2004) Degradation of polycyclic aromatic hydrocarbons in aqueous solutions by ultrasonic irradiation. *Journal of Hazardous Materials*. 108 (1–2), pp. 95–102.

Puls, R. W.; Blowes, D. W.; Gillham, R. W. (1999) Long-term performance monitoring for a permeable reactive barrier at the U.S. Coast Guard Support Center, Elizabeth City, North Carolina, *Journal of Hazardous Materials*, 68, pp. 109–124.

Puls, R.W.; Paul, C.J.; Powell, R.M. (1999) The application of in situ permeable reactive barrier technology for the remediation of chromate-contaminated groundwater: A field test. *Applied Geochemistry*, 14, pp. 989–1000.

Qiu, X.; Zhu, T.; Li, J.; Pan, H.; Li, Q.; Miao, G.; Gong, J. (2004) Organochlorine pesticides in the air around the Taihu Lake, China. *Environmental Science and Technology*, 38(5), pp. 1368–1374.

Råberg, S.; Nyström, M.; Erös, M.; Plantman, P. (2003) Impact of the herbicides 2,4-D and diuron on the metabolism of the coral porites cylindrical. *Marine Environmental Research*, 56, pp. 503–514.

Rahman, A.; Agrawal, A. (1997) Reduction of nitrate and nitrite by iron metal: Implications for ground water remediation. *Preprint Extended Abstract*, 213th ACS National Meeting, American Chemical Society, Division of Environmental Chemistry, 37(1), pp. 157–159.

Ramamoorthy, S. (1997) *Chlorinated organic compounds in the environment: Regulatory and monitoring assessment*, Lewis Publishers, Boca Raton, FL.

Ramesh, A.; Ravi, P.E. (2001) Applications of solid-phase microextraction (SPME) in the determination of residues of certain herbicides at trace levels in environmental samples. *Journal of Environmental Monitoring*, 3, pp. 505–508.

Rao, P.S.; Hayon, E. (1975) Oxidation of aromatic amines and diamines by OH radicals: Formation and ionization constants of amine cation radicals in water. *Journal of Physical Chemistry*, 79 (11), pp. 1063–1066.

Raupach, M.R.; Briggs, P.R.; Ford, P.W.; Leys, J.F.; Muschal, M.; Cooper, B.; Edge, V.E. (2001) Endosulfan transport: I. Integrative assessment of airborne and waterborne pathways. *Journal of Environmental Quality*, 30, pp. 714–728.

Redshaw, C.J. (1995) Ecotoxicological risk assessment of chemicals used in aquaculture: A regulation viewpoint. *Aquaculture Research*, 26, pp. 629–637.

Reeter, C. (1997) *Permeable reactive wall: Remediation of chlorinated hydrocarbons in ground water*. Technical Data Sheet, TDS-2047-ENV, Naval Facilities Engineering Service Center, Port Hueneme, CA.

Rivas, F.J.; Beltrán, F.J.; Frades, J.; Buxeda, P. (2001) Oxidation of *p*-hydroxybenzoic acid by Fenton's reagent. *Water Research*, 35, pp. 387–396.

Rochette, E.A.; Harsh, J.B.; Hill, H.H., Jr. (1993) Supercritical fluid extraction of 2,4-D from soils using derivatization and ionic modifiers. *Talanta*, 40(2), pp. 147–155.

Rock, M.L.; Kearney, P.C.; Helz, G.R. (1998) Innovative remediation technology. In P.C. Kearney, T. Roberts (eds.), *Pesticide remediation in soils and water*, John Wiley & Sons, New York.

Roe, B.A.; Lemley, A.T. (1997) Treatment of two insecticides in an electrochemical Fenton system. *Journal of Environmental Science and Health B*, 32(2), pp. 261–281.

Rose, A.J.; Waite, T.D. (2002) Kinetic model for Fe(II) oxidation in seawater in the absence and presence of natural organic matter. *Environmental Science and Technology*, 36, pp. 433–444.

Sakai, M. (2003) Investigation of pesticides in rainwater at Isogo Ward of Yokohama. *Journal of Health Science*, 49(3), pp. 221–225.

Santos, F.J.; Galceran, M.T.; Fraisse, D. (1996) Application of solid-phase microextraction to the analysis of volatile organic compounds in water. *Journal of Chromatography A*, 742, pp. 181–189.

Satterfield, C.N. (1991), *Heterogeneous catalysis in industrial practice*, 2nd ed., McGraw-Hill, New York.

Sawyer, D. (1991) *Oxygen chemistry*; Oxford University Press, New York.

Sayles, G.D.; You, G..; Wang, M.; Kupferle, M.J. (1997) DDT, DDD and DDE dechlorination by zero-valent iron. *Environmental Science and Technology*, 31, pp. 3448–3454.

Scherer, M.M.; Balko, B.A.; Tratnyek, P.G. (1998) The role of oxides in reduction reactions at the metal-water interface. In D.L. Sparks, T.J. Grudl (eds.), *Mineral-water interfacial reactions: kinetics and mechanisms*; ACS Symposium Series 715, American Chemical Society, Washington, DC, pp. 301–322.

Schneider, R.J. (1995) Evaluation of extraction method for triazine herbicides from soils for screening purposes. *Agribiological Research*, 48(3–4), pp. 193–206.

Schnoor, J.L.; Licht, L.A.; McCutcheon, S.C.; Wolfe, N.L.; Carriera, L.H. (1995) Phytoremediation of organic and nutrient contaminants. *Environmental Science and Technology*, 29, pp. 318–323.

Sedlak, D.L.; Andren, A.W. (1991) Aqueous-phase oxidation of polychlorinated biphenyls by hydroxyl radicals. *Environmental Science and Technology*, 25, pp. 1419–1427.

Seibig, S.; van Eldik , R. (1997) Kinetics of [FeII(edta)] Oxidation by molecular oxygen revisited. New evidence for a multistep mechanism. *Inorganic Chemistry*, 36, pp. 4115–4120.

Seibig, S.; van Eldik, R. (1999) [Multistep oxidation kinetics of [FeII(cdta)] [cdta = N,N',N'',N'''-(1,2-cyclohexanediamine)tetraacetate] with molecular oxygen. *European Journal of Inorganic Chemistry*, 3 pp. 447–454.

Sharma, V.; Millero, F.; Homonnay, Z. (2004) The kinetics of the complex formation between iron(III)-edta and hydrogen peroxide in aqueous solution. *Inorganic Chimica Acta*, 357, pp. 3583–3587.

Simandi, L. (1992) *Catalytic activation of dioxygen by metal complexes*; Kluwer Academic, Dordrecht, the Netherlands.

Simic, M.; Neta, P.; Hayon, E. (1971) Pulse radiolytic investigation of aliphatic amines in aqueous solution. *International Journal for Radiation Physics and Chemistry*, 3, pp. 309–320.

Sivakumar, M.; Gedanken, A. (2004) Insights into the sonochemical decomposition of $Fe(CO)_5$: Theoretical and experimental understanding of the role of molar concentration and power density on the reaction yield. *Ultrasonics Sonochemistry*, 71, pp. 373–378.

Smith, A.G.; Gangolli, S.D. (2002) Organochlorine chemicals in seafood: Occurrence and health concerns. *Food and Chemical Toxicology*, 40, pp. 767–779.

Solans, X.; Altaba, M.F. (1984) Crystal structures of ethylenediaminetetraacetato metal complexes. V. Structures containing the $[Fe(C_{10}H_{12}N_2O_8)(H_2O)]^-$ anion. *Acta Crystallographica*, C40, pp. 635–638.

Solans, X.; Altaba, M.F. (1985) Crystal structures of ethylenediaminetetraacetato metal complexes. VII. Structures containing the $[Fe(C_{10}H_{12}N_2O_8)(H_2O)]^-$ anion. *Acta Crystallographica*, C41, pp. 525–528.

Song, J.M. (2003) Oceanic iron fertilization: One of strategies for sequestration atmospheric CO_2. *Acta Oceanologica Sinica*, 22(1), pp. 57–68.

Sorokin, A.; Seris, J.; Meunier, B. (1995) Efficient oxidative dechlorination and aromatic ring cleavage of chlorinated phenols catalyzed by iron sulfophthalocyanine. *Science*, 268, pp. 1163–1166.

Sorokin, A.; Seris, J.; Meunier, B. (1996) Oxidation degradation of polychlorinated phenols catalyzed by metallosulfophthalocyanines. *Chemistry European Journal*, 2, pp. 1308–1317.

Stadtman, E.R.; Berlett, B.S. (1991) Fenton chemistry. Amino acid oxidation. *Journal of Biological Chemistry*, 266, pp. 17201–17211.

Stevenson, E.M. (1998) Incineration as a pesticide remediation method. In P.C. Kearney, T. Roberts (eds.), *Pesticide remediation in soils and water*, John Wiley & Sons, New York, pp. 85–103.

Stevenson, D.E.; Walborg, E.F., Jr.; North, D.W.; Sielken, R.L., Jr.; Ross, C.E.; Wright, A.S.; Xu, Y.; Kamendulis, L.M.; Klaunig, J.E. (1999) Monograph: reassessment of human cancer risk of aldrin/dieldrin. *Toxicology Letters*, 109, pp. 123–186.

Stipp, S.L.S; Hansen, M.; Kristensen, R.; Hochella, M.F., Jr.; Bennedsen, L.; Dideriksen, K.; Balic-Zunic, T.; Leonard, D.; Mathieu, H.-J. (2002) Behaviour of Fe-oxides relevant to contaminant uptake in the environment. *Chemical Geology*, 190, pp. 321–337.

Stravropoulos, P.; Celenligil-Cetin, R.; Tapper, A. (2001) The Gif Paradox. *Accounts of Chemical Research*, 34, pp. 745–752.

Stumm, W.; Morgan, J.J. (1996) *Aquatic chemistry: Chemical equilibria and rates in natural waters*, 3rd ed., John Wiley & Sons, New York, 780 pp.

Su, C.; Puls, R.W. (2001) Arsenate and arsenite removal by zero-valent iron: Kinetics, redox transformation, and implications for in situ groundwater remediation. *Environmental Science and Technology*, 35, pp. 1487–1492.

Sudo, M.; Kunimatsu, T.; Okubo, T. (2002) Concentration and loading of pesticide residues in Lake Biwa basin (Japan). *Water Research*, 36, pp. 315–329.

Sun, Y.; Pignatello, J.J. (1992) Chemical treatment of pesticide wastes-evaluation of Fe(III) chelates for catalytic hydrogen peroxide oxidation of 2,4-D at circumneutral pH. *Journal of Agricultural Food and Chemistry*, 40(2), pp. 322–327.

Suzuki, S.; Otani, T.; Iwasaki, S.; Ito, K.; Omura, H.; Tanaka, Y. (2003) Monitoring of 15 pesticides in rainwater in Utsunomiya, eastern Japan, 1999–2000. *Journal of Pesticide Science*, 28(1), pp. 1–7.

Sychev, A.Y.; Isak, V.G. (1995) Iron compounds and the mechanisms of the homogeneous catalysis of the activation of O_2 and H_2O_2 and of the oxidation of organic substrates. *Russian Chemical Reviews*, 64 (12), pp. 1105–1129.

Sychev, A.Y.; Isak, V.G.; Pfannmeller, U. (1979) Determination of rate constants of hydroxyl radicals with organic and inorganic substances under conditions for the catalytic decomposition of hydrogen peroxide. *Zhurnal Fizicheskoi Khimii*, 53 (11), pp. 2790–2793.

Tang, W.Z.; Chen, R.Z. (1996) Decolorization kinetics and mechanisms of commercial dyes by H_2O_2/iron powder system. *Chemosphere*, 32(5), pp. 947–958.

Tratnyek, P.G.; Johnson, T.L.; Schattauer, A. (1995) Interfacial phenomena affecting contaminant remediation with zero-valent iron metal. In P.C. Kearney, T. Roberts (eds.), *Emerging technologies in hazardous waste management VII*, John Wiley & Sons, New York, pp. 589–592. Preprint extended abstract presented before the I&EC Special Symposium, American Chemical Society, Atlanta, GA.

Tratnyek, P.G.; Scherer, M.M.; Johnson, T.J.; Matheson, L.J. (2003) Permeable reactive barriers of iron and other zero-valent metals. In M.A. Tarr (ed.), *Chemical degradation methods for wastes and pollutants: Environmental and industrial applications*, Marcel Dekker, New York, pp 371–421.

Troxler, W.L. (1998) Thermal desorption. In P.C. Kearney, T. Roberts (eds.), *Pesticide remediation in soils and water*, John Wiley & Sons, New York.

Tundo, P.; Anastas, P. (2000) *Green chemistry: Challenging perspectives*, Oxford University Press, New York.

Uhlig, H.H.; Revie, R.W. (1985) *Corrosion and corrosion control: An introduction to corrosion science and engineering*, 3rd ed., Wiley-Interscience, New York.

UNEP, ILO (1989) *Aldrin and dieldrin health and safety guide*, WHO, Geneva.

US Department of Energy Grand Junction Office (USDEGJO) (1989) Permeable reactive barriers. http://www.gjo.doe.gov/perm-barr/. Accessed in August 2004.

US Environment Protection Agency (2000) ChemicalWATCH factsheet. http://www.beyondpesticides.org. Accessed in August 2004.

US Environmental Protection Agency (2001) Breaking the cycle: 2001–2002 PBT program accomplishments. Chapter 1. http://www.epa.gov. Accessed in August 2004.

Van Veen, J.A.; van Overbeek, L.S.; van Elsas, J.D. (1997) Fates and activity of microorganisms introduced into soil. *Microbiology and Molecular Biology Reviews*, 61, pp. 121–135.

Vassilakis, C.; Pantidou, A.; Psillakis, E.; Kalogerakis, N.;Mantzavinos D. (2004) Sonolysis of natural phenolic compounds in aqueous solutions: Degradation pathways and biodegradability. *Water Research*, 38(13), pp. 3110–3118.

Vereen, D.A.; McCcall, J.P.; Butcher, D.J. (2000) Solid phase microextraction for the determination of volatile organics in the foliage of Fraser fir (*Abies fraser*). *Microchemical Journal*, 65, pp. 269–276.

Vidal, A.; Dinya, Z.; Mogyorodi, F.M., Jr.; Mogyorodi, F. (1999) Photocatalytic degradation of thiocarbamate herbicide active ingredients in water. *Applied Catalysis B: Environmental*, 21, pp. 259–267.

Voelker, B. (1994) *Iron redox cycling in surface waters: Effects of humic substances and light*. Doctor of Natural Science Dissertation, Diss. ETH No.10901, Swiss Federal Institute of Technology, Zurich.

Waite, T.D.; Joo, S.H.; Feitz, A.J.; Sedlak, D.S. (2004) Oxidative transformation of contaminants using nanoscale zero-valent iron. In H.H. Hahn, E. Hoffmann, H. Ødegaard (eds.), *Chemical water and wastewater treatment*, IWA Press, London, pp. 309–318.

Walling, C. (1975) Fenton's reagent revisited. *Accounts of Chemical Research*, 8, pp. 125–131.

Walling, C.; Kurz, M.; Schugar, H. (1970) The iron(III)-ethylenediaminetetraacetic acid-peroxide system. *Inorganic Chemistry*, 9, pp. 931–935.

Wan, C.; Chen, Y.H.; Wei, R. (1999) Dechlorination of chloromethanes on iron and palladium-iron bimetallic surface in aqueous systems, *Environmental Toxicology and Chemistry*, 18, pp. 1091–1096.

Wang, C.-B.; Zhang, W.-X. (1997) Synthesized nanoscale iron particles for rapid and complete dechlorination of TCE and PCBs. *Environmental Science and Technology*, 31(7), pp. 2154–2156.

Warren, K.D.; Arnold, R.G.; Bishop, T.L.; Lindholm, L.C.; Betterton, E.A. (1995) Kinetics and mechanism of reductive dehalogenation of carbon tetrachloride using zero-valent metals. *Journal of Hazardous Materials*, 41, pp. 217–227.

Watts, R.J.; Bottenberg, B.C.; Hess, T.F.; Jensen, M.D.; Teel, A.L. (1999) Role of reductants in the enhanced desorption and transformation of chloroaliphatic compounds by modified Fenton's reactions. *Environmental Science and Technology*, 33, pp. 3432–3437.

Watts, R.J.; Haller, D.R.; Jones, A.P.; Teel, A.L. (2000) A foundation for the risk-based treatment of gasoline-contaminated soils using modified Fenton's reactions. *Journal of Hazardous Materials*, 76, pp. 73–89.

Weber, J.B. (1994) Properties and behavior of pesticides in soil. In R.C. Honeycutt, D.J. Schabacker (eds.), *Mechanisms of pesticide movement into groundwater*, Lewis Publishers, London.

Weber, E.J. (1996) Iron-mediated reductive transformations: investigation of reaction mechanism. *Journal of Environmental Science and Technology*, 30, pp. 716–719.

Wells, C.F.; Salam, M.A. (1968) The effect of pH on the kinetics of the reaction of iron (II) with hydrogen peroxide in perchlorate media. *Journal of the Chemical Society [section] A:Inorganic, Physical, Theoretical*, 1, pp. 24–29.

Wenzel, W.W.; Adriano, D.C.; Salt, D.; Smith, R. (1999) Phytoremediation: a plant-microbe-based remediation system. In D.C. Adriano et al. (eds.), *Bioremediation of contaminated soils*, Agron. Monogr. 37, American Society of Agronomy, Madison, WI, pp. 457–508.

World Health Organization (1988) *Thiocarbamate pesticides: A general introduction*, World Health Organization, Geneva.

World Health Organization (1989) *Aldrin and dieldrin health and safety guide*, No. 21, World Health Organization, Geneva.

World Health Organization (1998) *Diazinon*, World Health Organization, Geneva.

Worthing, C.R.; Hance, R.J. (1991) *The pesticide manual*, 9th ed., The British Crop Protection Council, Surrey, U.K., pp. 243–244.

Wubs, H.; Beenackers, A.A.C.M. (1993) Kinetics of the oxidation of ferrous chelate of EDTA and HEDTA in aqueous solution. *Industrial and Engineering Chemistry Research*, 32, pp. 2580–2594.

Yabusaki, S.; Cantrell, K.; Sass, B.; Steefel, C. (2001) Multicomponent reactive transport in an in situ zero-valent iron cell. *Environmental Science and Technology*, 35, pp. 1493–1503.

Yak, H.K.; Wenclawiak, B.W.; Cheng, I.F.; Doyle, J.G.; Wai, C.M. (1999) Reductive dechlorination of polychlorinated biphenyls by zerovalent iron in subcritical water. *Environmental Science and Technology*, 33, pp. 1307–1310.

Yang, M.X.; Sarkar, S.; Bent, B.E. (1997) Degradation of multiply-chlorinated hydrocarbons on Cu(100). *Langmuir*, 13, 229–242.

Young, R.; Lopez-Avila, V. (1996) On-line determination of organochlorine pesticides in water by solid-phase microextraction and gas chromatography with electron capture detection. *Journal of High Resolution Chromatography*, 19(May), pp. 247–256.

Yurkova, I.; Schuchmann, H.; Sonntag, C. (1999) Production of OH radicals in the autoxidation of the Fe(II)-EDTA system. *Journal of the Chemical Society, Perkin Transactions 2*, pp. 2049–2052.

Zablotowicz, R.M.; Hoagland, R.E.; Locke, M.A. (1998) Biostimulation: Enhancement of cometabolic processes to remediate pesticide-contaminate soils. In P.C. Kearney, T. Roberts (eds.), *Pesticide remediation in soils and water*, John Wiley & Sons, New York.

Zang, V.; van Eldik, R. (1990) Kinetics and mechanism of the autoxidation of iron(II) induced through chelation by EDTA and related ligands. *Inorganic Chemistry*, 29, pp. 1705–1711.

Zećević, S.; Drazic, D.M.; Gojkovic, S. (1989) Oxygen reduction on iron: III. An analysis of the rotating disk-ring electrode measurements in near neutral solutions. *Journal of the Electrochemical Society*, 265(1–2), pp. 179–193.

Zećević, S.; Drazic, D.M.; Gojkovic, S. (1991) Oxygen reduction on iron-IV: The reduction of hydrogen peroxide as the intermediate in oxygen reduction reaction in alkaline solutions. *Electrochimica Acta*, 36(1), pp. 5–14.

Zhang, W.-X. (2003) Nanoscale iron particles for environmental remediation: An overview. *Journal of Nanoparticle Research*, 5, pp. 323–332.

Zhang, W-X.; Wang, C-B.; Lien, H-L. (1998) Treatment of chlorinated organic contaminants with nanoscale bimetallic particles, *Catalysis Today*, 40, pp. 387–395.

Zhou, X.; Mopper, K. (1990) Determination of photochemically produced hydroxyl radicals in seawater and freshwater. *Marine Chemistry*, 30, pp. 71–88.

Appendix A: XRD Analysis of ZVI Collected from Four Different Samples

(In the presence of molinate)

SAMPLE 7: ZVI (1.79 mM) + H_2O_2 (0.33 mM) in Milli-Q water
SAMPLE 8: ZVI in Milli-Q water
SAMPLE 9: ZVI in bicarbonate
SAMPLE 10: ZVI + H_2O_2 in bicarbonate

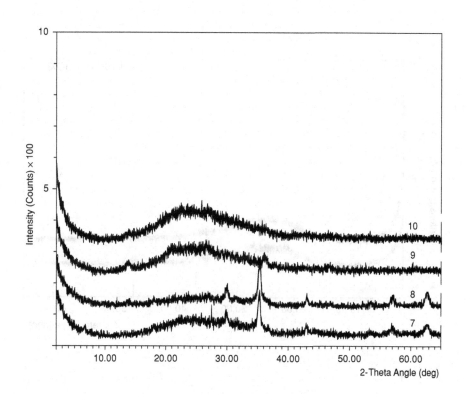

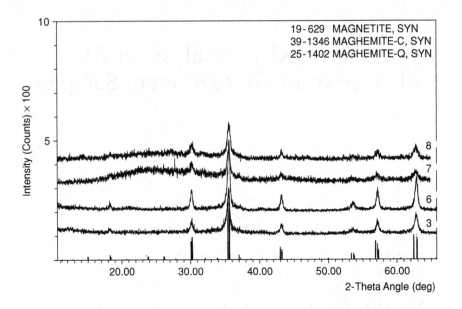

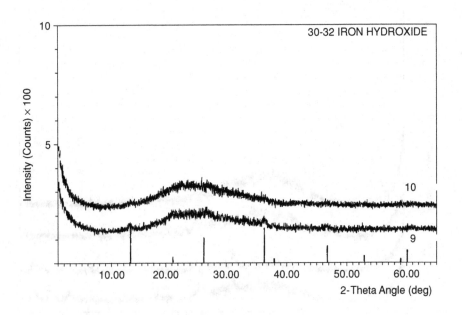

Appendix B: XRD Analysis of ZVI Collected from Four Different Samples

(In the absence of molinate)

Sample 1–4

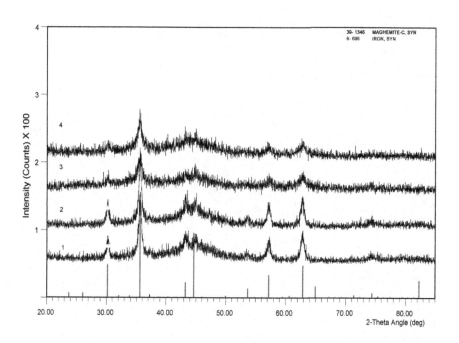

Appendix C: ELISA Analysis Methodology

Commercial enzyme-linked immunosorbent assay (ELISA) kits were used for molinate, chlorpyrifos, and diazinon analysis. Antibody and enzyme conjugates used for the cyclodiene insecticides and diuron analyses were provided by the Department of Agricultural Chemistry and Soil Science, University of Sydney.

A.1. A. Procedure for cyclodiene insecticides (e.g., endosulfan, endosulfan sulfate, aldrin, chlordane, dieldrin, heptachlor)

1. Dilute purified antibody stock (Rab $\propto$ ENDO II-KLH, 12.4 mg/mL) to 10 μg/mL in coating solutions; mix well but avoid frothing.
2. Add 100 μL to each well and incubate overnight at room temperature.
3. Wash three times with washing solution and tap dry on absorbent paper.
4. Add 200 μL of blocking solution (3% skim milk in phosphate buffer saline (PBS)) to each well and incubate for 1 h at room temperature.
5. Wash three times with washing solution and use immediately.
6. Prepare a standard curve.

A standard curve is obtained by plotting either the percentage of control absorbance or the percentage of inhibition of antibody binding or absorbance on y-axis in linear scale and the concentration of pesticide on x-axis in log scale (examples are shown below).

The percentage of control absorbance is determined by Equation (C.1). The percent inhibition is the inverse of the percentage of control and calculated using Equation (C.2).

$$\%B/B_0 = \frac{A - A_{\text{blank}}}{A_{\text{control}} - A_{\text{blank}}} \times 100 \tag{C.1}$$

$$\%\text{Inhibition} = \left[1 - \frac{A - A_{\text{blank}}}{A_{\text{control}} - A_{\text{blank}}}\right] \times 100 \tag{C.2}$$

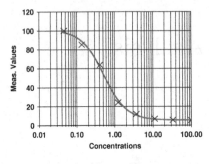

a. chlorpyrifos

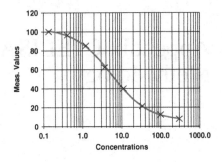

b. molinate

A = absorbance
$A_{control}$ = absorbance of pesticide at zero concentration
A_{blank} = absorbance of blank wells
$A_{control} - A_{blank}$ = absorbance of negative control
(The inhibition values range from 100% (absence of the hapten) to $n\%$ (hapten saturated antibodies.)

7. Prepare enzyme conjugate. For working stock, add 1 µL of stock enzyme conjugate (ENDO III-HRP; diuron-HRP) in 500 µL of 0.2% bovine serum albumin (BSA)-PBS.
8. For working solution, add 88 µL of working stock to 11 mL of 0.2% BSA-PBS (use this solution in the immunoassay). Enzyme conjugate solution should be prepared freshly in each experiment.
9. Bring all assay reagents including sample extracts to room temperature before use. Steps 10 and 11 should be performed within 15 min. Do not expose substrate to direct sunlight while pipetting or while incubating in the test wells.
10. Add 100 µL of standards and samples to their respective wells. Add 100 µL of purified water to control and blank wells.
11. Add 100 µL of enzyme conjugate to each well and thoroughly mix by moving in a circular motion on the benchtop. Incubate plate at room temperature for 1 h.
12. Wash six times with washing solution. Sit plates for 1–2 min between each wash. Tap dry on absorbent paper.
13. Mix 480 µL of substrate solution B and 15.52 mL of substrate solution A. Add 150 µL of substrate mix to each well.
14. Incubate plate at room temperature for 30 min. Blue color should develop.
15. Add 50 µL of stop solution to each well and mix thoroughly. This will turn the well contents yellow. Read as soon as possible (within 30 min) in a plate reader with a 450-nm filter. (*Note:* Before reading the microplates, ensure that there are no bubbles in any of the wells and no condensation on the bottom of the microplates. If necessary, rupture any bubbles with a clean pipette tip and wipe the bottom of the plate clean with a soft cloth.)

A.2. B. Procedures using commercial ELISA kits for the compounds of molinate, chlorpyrifos, and diazinon

B1. Molinate

A molinate plate kit (EnviroGard^{TM} Test Kit), which was purchased from Strategic Diagnostics Inc., was used for ELISA analysis of molinate. It contained a molinate stock solution (3 μg/mL), coated strip microwells, enzyme conjugate, substrate, and stop solution (13 mL of 1N HCl).

1. Prepare a standard curve.
2. Add 200 μL of blank and 100 μL of negative control (purified water), standards, and sample to separate wells.
3. Add 100 μL of enzyme conjugate to wells (except blank).
4. Mix contents in a circular motion and incubate at room temperature for 1 h
5. Empty contents into a sink and wash plate under cool running water five times.
6. Add 100 μL gof substrate using a multichannel pipette and mix in a circular motion, before incubating at room temperature for 30 min.
7. Add stop solution of 100 μL (1 N HCl) using a multichannel pipette and mix in a circular motion.
8. Read developed color as soon as possible (within 30 min) in a plate reader at the absorbance of 650–450 nm.

B2. Chlorpyrifos

The chlorpyrifos plate kit (EnviroGard^{TM} Test Kit), purchased from Strategic Diagnostics Inc., contained a chlorpyrifos stock solution (1 μg/mL), coated strip microwells, enzyme conjugate, substrate, and stop solution (13 mL of 1 N HCl).
Analytical procedures are the same as for molinate (B1).

B3. Diazinon

The diazinon plate kit (EnviroGard^{TM} Test Kit), also purchased from Strategic Diagnostics Inc., contained diazinon stock solution (100 ng/mL), coated strip microwells, enzyme conjugate, substrate, and stop solution (1 N HCl). Analytical procedures are the same as for molinate (B1).

Appendix D: Oven Programs for GC/MS Analysis of Pesticides

Oven ramp	°C/min	Next (°C)	Hold (min)	Run time
		Atrazine		
Initial		50	1	1
Ramp 1	15	150	0.5	8.17
Ramp 2	4	180 (210)	1	16.67 (24.17)
Ramp 3	0			
Postrun				16.67 (24.17)
		Molinate (for quantitative analysis)[a]		
Initial		80	1	1
Ramp 1	30	178	3	7.27
Ramp 2	30	250	5	14.67
Ramp 3	0			
Postrun				14.67
		Molinate (for qualitative analysis)		
Initial		80	0	0
Ramp 1	5	200	0	24
Ramp 2	5	210	3	29
Ramp 3	20	270	3	35
Post run				35
		Endosulfan		
Initial		140	1	1
Ramp 1	30	180	0	2.33
Ramp 2	10	250	6	15.33
Ramp 3				
Postrun				15.33
		Chlorpyrifos		
Initial		50	1	1
Ramp 1	15	150	0.5	8.17
Ramp 2	4	210	1	24.17
Ramp 3				
Postrun				24.17

[a] To achieve satisfactory limits of detection of target compounds, the SIM mode was used.

- Inlet: injection volume – 2μL; inlet front – EPC split-splitless inlet; mode – splitless; gas – He; heater (°C) – 260; pressure (psi) – 10.7 (atrazine, chlorpyrifos), 10.4 (endosulfan, chlorpyrifos), 9.3 (molinate), 12.8 (endosulfan); total flow (mL/min) – 4.5 (atrazine, chlorpyrifos), 4.2 (molinate, endosulfan); purge flow to split vent – 1mL/min@0.9 min.
- Washes: Pre-injection – 2 times in sample, 3 times in solvents, 3 times in pumps; postinjection – 4 times in solvents.
- Mode: Const flow, He flow – pressure 10.8 psi (atrazine, chlorpyrifos), 9.4 psi (molinate), 12.8 psi (endosulfan); flow – 1.3 mL/min (atrazine, chlorpyrifos), 1mL/min (molinate, endosulfan); average velocity – 42cm/s (atrazine, chlorpyrifos), 37cm/sec (molinate), 38 cm/s (endosulfan); installed column – inventory #, 29.8m × 250μm × 0.25μm calibrated.
- Manufacturer's specifications: Model no. – HP 19091S – 433, 325°C max, HP – 5MS; 5% phenyl methyl siloxane, capillary; 30m × 250μm × 0.25μm nominal, aux; heater – 280 °C; oven configuration – maximum (300°C); equilibrium (0.5 min).

Appendix E: Experimental Conditions for Pesticides and Preliminary Screening Studies Using Nanoscale ZVI

Experiments were conducted using two analytical techniques, SPME GC/MS and ELISA. Samples in ELISA methods were analyzed without filtration as this enabled confirmation that the agrochemicals were indeed degraded and not removed from solution by adsorption. Controls were undertaken to confirm that ELISA tests were unaffected by the presence of the nZVI particles, as shown in the following figure.

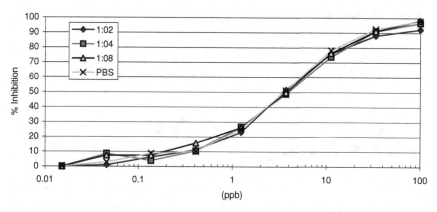

Big Wig and its sensitivity to Endo I-HRP(10)

FIGURE E.1. Test of ZVI particles for the inhibition from antibody in ELISA analysis (concentration of ZVI particles: 1.8 mM).

TABLE E.1 Experimental conditions for pesticides.

Compounds	Analysis	Initial concentration (μg/L)	ZVI (mM)	Other conditions	pH
α -, β- Endosulfan	ELISA	10	8.9	O_2, NF	6.8
Endosulfan sulfate	ELISA	50 [1] 10 [2]	1 [1] 8.9 [2]	O_2, NF [1] O_2, NF [2]	4 [1] 4, 6.8 [2]
	GC/MS	50 [1]	1 [1]	N_2, F [1]	4 [1]
Aldrin	ELISA	20 [3] 10 [4]	5.36 [3] 8.9 [4]	O_2, after 3 h NF [3] O_2, After 5 h NF [4]	6.8 [3] 6.8 [4]
Dieldrin, heptachlor, chlordane	ELISA	10	8.9	O_2, NF	6.8
Atrazine	GC/MS	1000	8.9, 35.7	N_2, F	6.8
Molinate	ELISA [5]	10	31.3	O_2, NF	—
	GC/MS [6]	100	10.7	O_2, F	—
Chlorpyrifos	ELISA	10	25	O_2, NF	—
Diazinon	ELISA	0.1	5.36, 10.7	O_2, NF	—
Diuron	ELISA	100	17.9	O_2, NF	6.8

Note: pH 4 was adjusted with 10^{-4} M HCl, while pH 6.8 was adjusted with 0.02 M phosphate buffered solution. NF: No filtering; F: filtering.

Results of pesticides

(1) Endosulfan sulfate

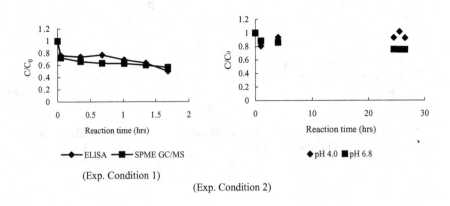

(Exp. Condition 1)

(Exp. Condition 2)

(2) α-, β- Endosulfan

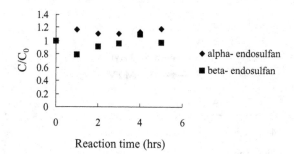

(3) Aldrin

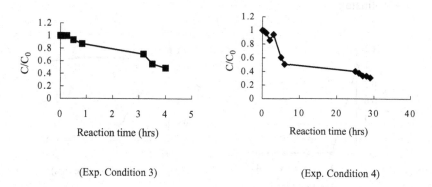

(Exp. Condition 3)   (Exp. Condition 4)

(4) Dieldrin, Chlordane

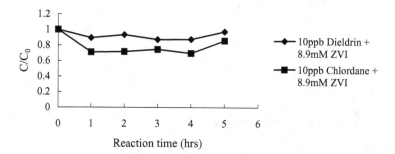

(5) Atrazine

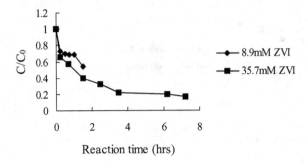

(6) Molinate

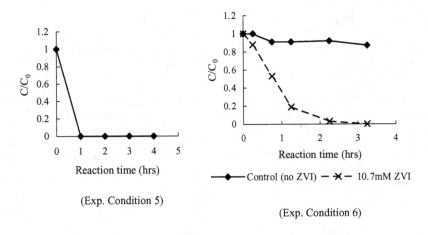

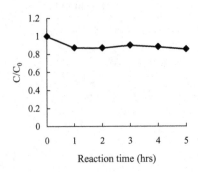

(Exp. Condition 5)

(Exp. Condition 6)

(7) Diuron

(8) Chlorpyrifos

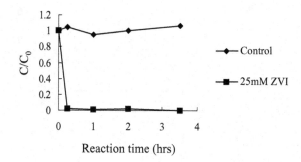

(9) Diazinon

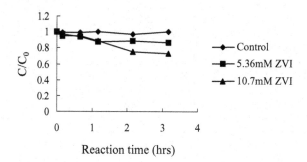

Index